DEVELOPMENTAL GENOMICS OF ASCIDIANS

DEVELOPMENTAL GENOMICS OF ASCIDIANS

NORIYUKI SATOH

WILEY Blackwell

Wiley Blackwell is an imprint of John Wiley & Sons, formed by the merger of Wiley's global Scientific Technical and Medical business with Blackwell Publishing.

Published by John Wiley & Sons, Inc., Hoboken, New Jersey.
Published simultaneously in Canada.

For general information on our other products and services or for technical support, please contact our Customer Care Department within the United States at (800) 762-2974, outside the United States at (317) 572-3993 or fax (317) 572-4002.

Wiley also publishes its books in a variety of electronic formats. Some content that appears in print may not be available in electronic formats. For more information about Wiley products, visit our web site at www.wiley.com.

Library of Congress Cataloging-in-Publication Data:

Satoh, Noriyuki.
 Developmental genomics of ascidians / by Noriyuki Satoh.
  1 online resource.
 Description based on print version record and CIP data provided by publisher; resource not viewed.
 ISBN 978-1-118-65604-4 (Adobe PDF) – ISBN 978-1-118-65624-2 (ePub) – ISBN 978-1-118-65618-1 (cloth) 1. Sea squirts–Anatomy. 2. Sea squirts–Development. I. Title.
  QL613
  596′.2–dc23
                                        2013038006

Printed in Singapore

10 9 8 7 6 5 4 3 2 1

*To all who have contributed
to the developmental biology of ascidians*

CONTENTS

PREFACE · ix

CHAPTER 1 · *A BRIEF INTRODUCTION TO ASCIDIANS* · 1

CHAPTER 2 · *THE DEVELOPMENT OF TADPOLE LARVAE AND SESSILE JUVENILES* · 9

CHAPTER 3 · *GENOMICS, TRANSCRIPTOMICS, AND PROTEOMICS* · 19

CHAPTER 4 · *RESEARCH TOOLS* · 31

CHAPTER 5 · *THE FUNCTION AND REGULATION OF MATERNAL TRANSCRIPTS* · 41

CHAPTER 6 · *LARVAL TAIL MUSCLE* · 53

CHAPTER 7 · *ENDODERM* · 63

CHAPTER 8 · *EPIDERMIS* · 69

CHAPTER 9 · *NOTOCHORD* · 77

CHAPTER 10 · *THE LARVAL AND ADULT NERVOUS SYSTEMS* · 89

CHAPTER 11 · *MESENCHYME* · 107

CHAPTER 12 · *MAKING BLUEPRINT OF CHORDATE BODY: DYNAMIC ACTIVITIES OF REGULATORY GENES* · 113

CHAPTER 13 · *DEVELOPMENT OF THE JUVENILE HEART* · 137

CHAPTER 14 · *GERM-CELL LINE, GAMETES, FERTILIZATION, AND METAMORPHOSIS* · 145

CHAPTER 15 · *INNATE IMMUNE SYSTEM AND BLOOD CELLS* · 159

CHAPTER 16 · *COLONIAL ASCIDIANS: ASEXUAL REPRODUCTION AND COLONY SPECIFICITY* · 167

CHAPTER 17 · *EVOLUTIONARY DEVELOPMENTAL GENOMICS* · 175

INDEX · 193

PREFACE

I have been captivated by ascidian embryology for more than 35 years, starting with *Halocynthia roretzi* and followed by *Ciona intestinalis*. There are numerous reasons why I still work with ascidians after all this time. The first and simplest reason is that ascidian embryogenesis, both temporally and spatially, is well suited to my individual research interests. While a developmental biologist's personality impacts his/her choice of experimental system, the model organisms themselves often dictate the research topics or themes. For example, vertebrates develop slowly, whereas ascidians do so quickly. Longer developmental time periods lend themselves to mechanistic investigations of detailed developmental processes; however, it is difficult for a vertebrate embryologist to follow the entire process of embryogenesis in a single study. By contrast, because I am interested in understanding the clock mechanisms that control the timing of developmental events, I selected *Halocynthia* as my first model organism because of its comparatively short period of embryogenesis. As I began studying the timing of differentiation marker gene expression, the length of *Halocynthia* embryogenesis was very reasonable—development from fertilization to the newly formed larva stage took approximately one and half days at 15 °C. Even better, in *Ciona*, the fertilized egg developed to the larva within 18 hours at 18 °C. This time frame of ascidian development has affected my research outlook to the point that I am less concerned with individual developmental processes, and prefer to view and investigate embryogenesis as a whole.

The second reason for my fidelity to ascidians comes from their beautiful embryos and invariant cleavage patterns. Under a stereomicroscope, we observe well-regulated cleavage, gastrulation, neurulation, and tailbud embryo formation. We can identify every blastomere up to gastrulation. All processes are very dynamic, with fewer constituent cells than vertebrate embryos. In addition, the formed tadpole larva represents the most simplified body plan of the chordates, a group to which humans also belong.

Soon after I started my research on ascidian embryos, I became convinced that ascidians would become one of the most suitable experimental systems in which to explore the cellular and molecular mechanisms of embryonic cell specification and differentiation. These were the main reasons why I published the book *Developmental Biology of Ascidians*, by Cambridge University Press in 1994. That book aimed to review extant ascidian embryology research, as well as highlight the advantages of the model system and encourage more people to join the field.

A dramatic change in ascidian embryology occurred in 2002, when the draft genome of *C. intestinalis* was sequenced after a collaborative effort of worldwide ascidian developmental biologists with the Joint Genome Institute, Department of Energy, USA. This project identified almost all of the developmentally relevant genes in the *Ciona* genome, which represents the basic set of gene components in chordates prior to the two genome-wide gene duplications that occurred in vertebrates. The absence of redundancy in ascidian regulatory gene functions makes it easier to elucidate the developmental roles of individual genes. In association with the genome sequencing project, a cDNA sequencing

ix

project was also carried out, providing great quantity of information on the expression profiles of regulatory genes. Whole-mount *in situ* hybridization reveals distinct gene expression profiles in embryos at the single cell level. We, ascidian embryologists, are proud that we can discuss mechanisms at the single cell level, rather than at the regional embryonic level that is usually discussed in vertebrate embryos.

By applying new molecular techniques to the ascidian system, we have answered questions posed by several pioneer ascidian embryologists. We now know the gene for the muscle determinant activity, first identified by Conklin. We now know the molecular mechanisms involved in self- and nonself-recognition of gametes of hermaphroditic *C. intestinalis*. Morgan investigated this question several times. We now know the molecular mechanisms involved in ascidian embryo notochord formation, a characteristic feature of chordates. Kowlevsky first pointed out the enormous opportunity afforded by ascidians to bridge the big evolutionary gap between invertebrates and vertebrates. The simplicity of the ascidian embryology and genome, with its basic set of chordate regulatory genes, together with the modern techniques available to researchers—including forward genetics—have made the *Ciona* embryo one of the best systems in which to study developmental genomics. I am confident that this system will continue to develop quickly in the years to come.

As one of the comparatively old fellows studying ascidian embryology, for me one of the greatest pleasures is to see newcomers to the field who have already enjoyed great research careers using other model systems, for example, Michael Levine and Patrick Lemaire. These established scientists are accompanied by young researchers with astonishing productivity. Most of the current, elegant experiments in the field have been carried out by these young ascidian embryologists, who are a real treasure for the community. In addition, these scientists have published good reviews on the current progress in their respective subsets of ascidian developmental genomics. I have cited their work extensively in this book. I appreciate deeply their kind consent for me to use their descriptions, and for their comments and criticisms during the draft preparation stage of this book.

Special thanks go to Drs. Hiroki Nishida, Yutaka Satou, Patrick Lemaire, Micheal Levine, Clare Hudson, Hitoyoshi Yasuo, Lionel Christiaen, Kohji Hotta, Yasunori Sasakura, Kaoru S. Imai, Brad Davidson, William R. Jeffery, Billie Swalla, William C. Smith, Di Jiang, Kazuo Kawamura, Shirae-Kurabayashi, Anthony De Tomaso, Alberto Stolfi, Shigehiro Yamada, Hiroki Takahashi, Yoshito Harada, Rixy Yamada, Hitoshi Sawada, Miho Suzuki, Robert Zeller, Keisuke Nakashima, Teruaki Nishikawa, Maki T. Kobayashi, Hidetoshi Saiga, Takehiro Kusakabe, Naohito Takatori, Honoo Satake, Masaru Nonaka, Tatsuya Ueki, Hitoshi Michibata, Eiichi Shoguchi, Takeshi Kawashima, Fuki Gyoja, Christian Sardet, Francois Prodon, Arend Sidow, Anna Di Gregorio, Jean-Stepane Joly, Kazuo Inaba, Michio Ogasawara, Kasumi Yagi, and Daniel Rokhsar.

In one sense, this book is a composite of review articles contributed by many experts in the field. However, it also contains my thoughts on ascidian embryogenesis as viewed from a long perspective. What I especially want to convey in this book is the urgent need to describe a complete (or nearly complete) set of molecular and cellular events associated with *Ciona* embryogenesis so that we can address the final question of embryogenesis. Specifically, with reference to Lewis Wolpert's "It is not birth, marriage, or death, but gastrulation which is truly the most important time in your life," my response is "Embryogenesis is not a simple series of changes to embryonic cells, but rather the place to develop phenotypic individuality as an extant organism with a long evolutionary history."

We can begin to understand our development as individuals and as a species through studies of ascidian embryogenesis.

Finally, I would like to thank the members of the Molecular Developmental Biology Laboratory at Kyoto University. Kazuko Hirayama is acknowledged for her great support at Kyoto University. Thanks to Kanako Hisata for her great help in preparing the figures and tables and Shoko Yamakawa for typing the manuscript. I am especially grateful to my wife, Mikako Satoh, for her daily support of my research.

The title of this book, *Developmental Genomics of Ascidians*, reflects my desire for future development of this field using ascidian embryos. If readers of this book come away impressed by and attracted to the ascidians, especially the *Ciona* system, it is my unbidden pleasure. If they come away uninspired by the *Ciona* system, it is simply due to my inability to describe it adequately. At any rate, I am happy to leave this book as my last and largest contribution to ascidian developmental biology.

NORIYUKI SATOH

Okinawa, Japan

CHAPTER 1

A BRIEF INTRODUCTION TO ASCIDIANS

1.1 WHAT ARE ASCIDIANS?

Ascidians, or sea squirts, are sessile marine invertebrate chordates ubiquitous throughout the world. The name "ascidian" originated from the Greek word *askidion*, meaning a small bag or vase. The class comprises approximately 2900 extant species,[1] most of which live in shallow water. Ascidians usually attach to rocks, shells, and pilings, and live by filtering tiny plankton and other nutrients from seawater. The entire adult body is invested with a thick covering, the tunic (or test), from which the subphylum name, Tunicata is derived. A major constituent of the tunic is tunicin, a type of cellulose. Tunicates or urochordates are the only animals that can synthesize cellulose independently.

An individual ascidian has two openings, an incurrent oral (branchial) siphon and an outcurrent atrial siphon (Fig. 1.1a, b). The mouth behind the oral siphon leads to a large pharynx, or branchial basket—a chamber perforated by dorsoventral rows of numerous gill slits called stigmata (Fig. 1.1b). Along the ventral margin of the branchial basket is a specialized organ called the endostyle, which secretes large quantities of mucus used for capturing food particles (Fig. 1.1b). The endostyle contains iodine, and therefore this organ has an evolutionary relationship with the vertebrate thyroid gland. The digestive tract leads to a stomach at the bottom of the U-shaped digestive loop, followed by an intestine that terminates at the anus, which opens into the atrial cavity (Fig. 1.1b).

The adult nervous system consists of a single cerebral ganglion lying between the two siphons and an adjacent neural gland (Fig. 1.1b). Several nerves elongate from the ganglion to various parts of the body, including the muscles, pharynx, viscera, gonad, and siphons (cover picture). By contrast, the neural gland leads through a duct to the pharynx, just behind the mouth. The open circulatory system is well developed and consists of a short, tubular heart and numerous blood vessels. The heart lies posteroventrally in the body near the stomach and behind the pharyngeal basket (Fig. 1.1b, o). The heartbeat and the direction of blood flow reverse periodically. The circulatory system contains several different types of blood cells or coelomic cells with specialized functions.

[1]Based on a recent estimation of Appeltans *et al.* (2012) The magnitude of global marine species diversity. *Curr. Biol.*, 22, 2189–2202.

Developmental Genomics of Ascidians, First Edition. Noriyuki Satoh.
© 2014 John Wiley & Sons, Inc. Published 2014 by John Wiley & Sons, Inc.

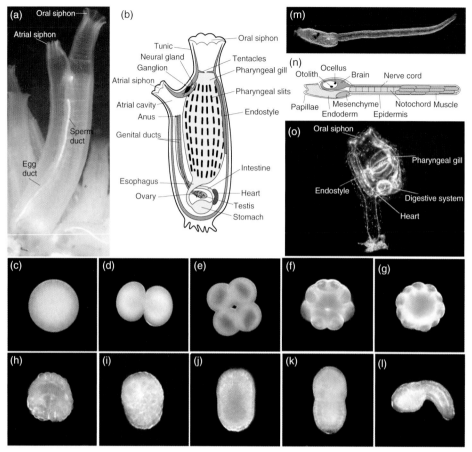

FIGURE 1.1   The tunicate ascidian *Ciona intestinalis*. (a) An adult with oral (incurrent) and atrial (outcurrent) siphons. The white duct is the sperm duct and the orange duct paralleling it is the egg duct. (b) Diagram illustrating adult organs and tissues. (c–l) Embryogenesis. Embryos were dechorionated to clearly show their outer morphology. (c) Fertilized egg, (d) 2-cell embryo, (e) 4-cell embryo, (f) 16-cell embryo, (g) 32-cell embryo, (h) gastrula (~150 cells), (i, j) neurulae, (k, l) tailbud embryos, and (m) tadpole larva. (n) Diagram illustrating larval organs and tissues. (o) A juvenile a few days after metamorphosis, with internal structures labeled.

1.1.1   Taxonomy

Ascidians belong to the class Ascidiacea, subphylum Urochordata (Tunicata), and phylum Chordata (Fig. 1.2). Animals with a notochord or a rod-shaped axial organ and a dorsal hollow neural tube are classified into the phylum Chordata. Chordata consists of three subphyla, Cephalochordata, Urochordata, and Vertebrata (Fig. 1.2). Cephalochordates (lancelets or amphioxus) are headless (acraniates), and have well-segmented somites and a notochord running throughout the body. Urochordates contain a notochord in the tail during at least the larval stage, and have a well-organized larval central nervous system. Vertebrates or craniates develop vertebrae from the notochord, along with jaws, heads, and an adaptive immune system. Chordates were present on the Earth at least ~520 million years (MYR) ago, as demonstrated by recent fossil records of Chenjan Fauna. The origin

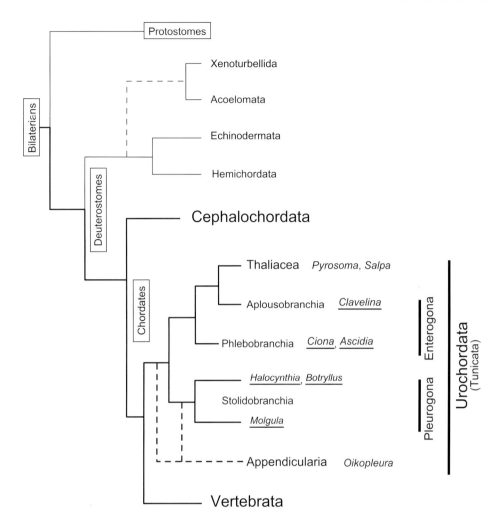

FIGURE 1.2   A phylogenetic tree indicating the position of urochordates (tunicates) among deuterostomes and chordates. Several ascidians species (underlined) are listed in the corresponding phylogenetic positions. The phylogenetic relationship of Appendicularia (larvaceans) still remains to be determined (bold dotted lines).

and evolution of chordates have been debated for more than 150 years. Recent molecular phylogeny and comparative genomics studies demonstrated that cephalochordates represent the most ancient and extant chordate lineage. Furthermore, urochordates and vertebrates were shown to form a sister group (Olfactants), indicating that urochordates are the invertebrates most closely related to vertebrates (Fig. 1.2).

The subphylum Urochordata or Tunicata is composed of three classes, Ascidiacea (ascidians), Thaliacea (salps), and Appendicularia (larvaceans) (Fig. 1.2). The evolutionary relationships among tunicates remain controversial because of their great variety of life cycles and rapid evolution. The class Ascidiacea comprises two subclasses, Enterogona and Pleurogona (Fig. 1.2). The Enterogona ascidians are characterized by the location of the gonad (a single ovary and a single testis; ascidians usually are hermaphrodites) within the gut loop or posterior to it. This subclass consists of Aplousobranchia (*Clavelinai* and others) and Phlebobranchia (*Ciona intestinalis, Phallusia mammillata*, and others) (Fig. 1.2). On the other hand, the Pleurogona ascidians, which have a pair of gonads inside the right and left body walls, include Molgulidae (*Molgula oculuta* and others), Styelidae (*Styela plicata* and others), and Pyuridae (*Halocynthia roretzi, Botryllus schlosseri*, and others) (Fig. 1.2). Some of these organisms live as individuals (solitary or simple ascidians), while others form colonies (colonial or compound ascidians). The colonial life style evolved several times independently in the orders.

1.1.2 Reproduction

Ascidians are hermaphrodites (Fig. 1.1a, b). Sexual reproduction is common in solitary ascidians, whereas colonial ascidians reproduce both sexually and asexually by budding. Colonial ascidians also have an extensive capacity for regeneration. Stereotypic embryogenesis of solitary ascidians proceeds rapidly (Fig. 1.1c–l). Bilaterally symmetrical cleavage takes place according to a highly determinate pattern (Fig. 1.1d–g). Gastrulation begins around the 120-cell stage and is followed by neurulation (Fig. 1.1h–j). Then, tailbud embryos are formed (Fig. 1.1k, l), and finally a conventional tadpole-type larva hatches from the chorion (Fig. 1.1m), usually within 12 h to a few days after fertilization. The ascidian tadpole consists of only ∼2600 cells, but has distinct tissues and organs, including the epidermis, central and peripheral nervous systems, endoderm (which gives rise to the adult digestive tract and its associated organs), notochord, muscle, and mesenchyme (from which several adult mesodermal organs are derived) (Fig. 1.1n). These organs represent the major components of any vertebrate body, including our own, and the mechanisms underlying their formation are the focal point of this book.

The ascidian tadpole larva differs from the amphibian tadpole in that it has no mouth. The nonfeeding larva swims for a few hours or more as solitary ascidians, and then metamorphoses into a juvenile (Fig. 1.1o). The growth of the juvenile is also rapid. For example, eggs of the most common and cosmopolitan species, *C. intestinalis*,[2] give rise to adults with reproductive capacity within 2–3 months, or earlier in warm waters, suggesting that under optimal conditions *Ciona* can pass through several generations within a year. Closed-system inland culture conditions have been established for the maintenance of these organisms for laboratory study. In Japan, *C. intestinalis* are provided to researchers all year round through support of the National BioResource Project (NBRP).

1.2 A BRIEF HISTORY OF RESEARCH ON ASCIDIAN EMBRYOS

Ascidians have been recognized since the ancient Greeks, and were described by Aristotle. Because of their soft bodies, they were long classified as a group of mollusks by Carl Linnaeus. At that time, there was a noticeable gap between vertebrates and invertebrates. After the publication of Charles Darwin's "On the Origin of Species by Means of Natural Selection" in 1859, evolutionary theories came into vogue to describe relationships

---

[2]Recent studies demonstrate that there are at least two cryptic species of *C. intestinalis*, one (Sp. A) distributing worldwide and the other (Sp. B) mainly from North Atlantic [e.g., Caputi *et al.* (2007) Cryptic speciation in a model in vertebrate chordate. *PNAS*, 104, 9364–9369]. Description of *C. intestinalis* in this book is of Sp. A, otherwise denoted.

among animals. In 1886, the great Russian embryologist Alexander Kowalevsky (Fig. 1.3a) discovered that the ascidian larva has the general appearance of a simplified vertebrate tadpole (Fig. 1.3b). This observation, together with his other finding that amphioxus possess a notochord, positioned the ascidians and amphioxus as a new group, the protochordates, which were intermediate between invertebrates and vertebrates—filling the evolutionary gap. More recently, it seems that the long debate on the origin and evolution of chordates has reached a consensus opinion that cephalochordates are basal among chordates, while urochordate ascidians are a sister group to the vertebrates.

It is said that the descriptive work of van Beneden and Julin in 1884 was the first demonstration of the relationship between the egg axis and the larval body plan. The door to classical experimental embryology was opened in 1887 by the French anatomist, Laurent Chabry (Fig. 1.3c), in a study of ascidian embryos. Chabry destroyed one blastomere of a 2-cell *Ascidiella aspersa* embryo and found that the remaining blastomere continued to cleave as if it were half of the whole embryo, eventually forming a half-larva (instead of a complete dwarf larva) (Fig. 1.3d). After obtaining a similar result using a 4-cell embryo, he inferred that ascidian embryos could not compensate for missing parts, and thus, that the developmental pattern was "mosaic." His experiment was followed by Roux's studies of amphibian embryos in 1888 and Driesch's work on sea urchin embryos in 1891. The triumph in this vein of experimental embryology was, of course, the 1924 discovery by Spemann and Mangold of the "organizer," as determined by transplantation experiments on amphibian embryos.

The pioneering zoologist and geneticist, Thomas H. Morgan, carried out many seminal studies not only in *Drosophila* but also in other animals, including ascidians. He was especially puzzled by the self-sterility of *C. intestinalis*, which he first reported in 1904. Ascidians are hermaphrodites and self-fertilization is usually blocked. Morgan published ten more reports on this topic until 1945. His attempt to understand the molecular mechanisms of self-sterility has now been realized by young Japanese scientists taking full advantage of genetics, genomics, and proteomic techniques available for *C. intestinalis*. The molecular mechanisms involved in the ascidian self and nonself recognition are likely to be similar to those that have been recently unveiled in plants.

In 1905, in a milestone of ascidian embryology, Edwin G. Conklin (Fig. 1.3e) described the lineage of embryonic cells and the segregation of egg cytoplasmic regions into specific larval organs, suggesting that maternal factors were responsible for the later differentiation of embryonic cells (Fig. 1.3f). He found that the lineage of embryonic cells is invariant, and that most of the embryonic cells are destined to give rise to one type of tissue or organ during cleavage stages prior to the initiation of gastrulation. Conklin's insight was followed by a considerable number of experimental studies on ascidian embryogenesis, carried out between 1940–1960, particularly by Italian embryologists, including Jossepp Reverveli, Giuseppina Ortolani, and others.

In 1973, Richard Whittaker, using cleavage-arrested embryos, showed that maternally provided factors are segregated into specific lineages of embryonic cells, and that these factors trigger developmental pathways leading to cell differentiation, as Conklin proposed. Whittaker's study initiated the field of molecular developmental biology, which incorporated techniques, rules, and concepts from molecular biology. Ascidians were proposed as a model experimental system for molecular developmental biology, particularly with respect to the study of mechanisms underlying embryonic cell fate specification. This new research focus was the main impetus behind the 1994 publication of *Developmental Biology of Ascidians*.

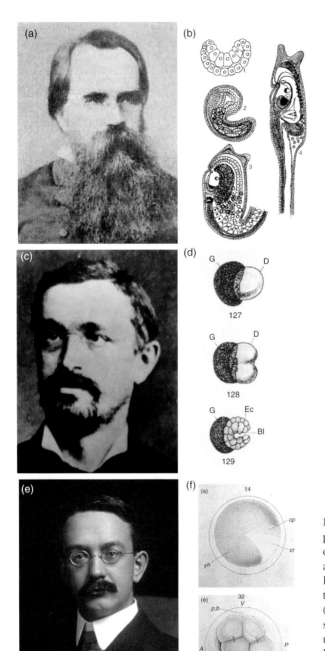

FIGURE 1.3 Three great pioneers of ascidian embryology with their descriptions of ascidian embryos. (a) Alexander Kowalevsky and (b) his illustrations of ascidian embryos. (c) Lawrent Chabry and (d) his sketches depicting the manipulation of *Ascidiella* eggs. (e) Edwin G. Conklin and (f) his drawings of ascidian embryos. (Photographs were obtained from Wikipedia.)

A big turn in ascidian embryology came in 2002, when the draft genome of *C. intestinalis* was published. At that time, it was the seventh animal with sequenced genome. Importantly, the *C. intestinalis* draft genome revealed that the ascidian contains a basic set of developmentally relevant genes that are also used by vertebrates, including transcription factors (TFs) and cell–cell signaling pathway molecules (SPMs). Thus, *Ciona* continues to be an appropriate experimental system for a diverse range of biological studies.

The recent advances in the field of ascidian embryology are extraordinary. Owing to the well-described lineages, embryonic cell specification mechanisms are now understandable in view of "gene regulatory networks." Imaging of embryos and juveniles has greatly advanced our understanding of the constitution of the ascidian embryo at the single cell level. Forward and reverse genetics have revealed the functions of developmental regulatory genes. With embryologic and genomic simplicity, *Ciona* is a model system for developmental studies of chordates. The main aim of this book is to introduce their attractiveness for future mechanistic studies.

SELECTED REFERENCES

BRUSCA, R.C., BRUSCA, G.J. (2003) The urochodates (tunicates), in *Invertebrates*, 2nd edn, Sinauer Assoc. Inc, Sunderland, MA, pp. 855–864.

CHABRY, L. (1887) Contribution a l'embryologie normale et teratogique des Ascidies simples. *J Anat Physiol (Paris)*, **23**, 167–319.

CLONEY, R.A. (1990) Urochordata-Ascidiacea, in *Reproductive Biology of Inverterbates*, (eds K.G. Adiyodi and R.G. Adiyodi), Oxford and IBH, New Delhi, pp. 361–451.

CONKLIN, E.G. (1905) The organization and cell lineage of the ascidian egg. *J Acad Natl Sci (Philadelphia)*, **13**, 1–119.

JEFFERY, W.R., SWALLA, B.J. (1997) Tunicates, in *Embryology*, (eds S.F. Gilbert and A.M. Raunio), Sinauer, Sunderland, MA, pp. 331–364.

KOWALEVSKY A (1866) Entwicklungsgeschichte der einfachen Ascidien. *Memory l'Academy St Petersbourg*, **7** (10), 1–19.

MORGAN, T.H. (1944) The genetic and the physiological problems of self-sterility in *Ciona*. VI. Theoretical discussion of genetic data. *J Exp Zool*, **95**, 37–59.

REVERBERI, G. (1971) Ascidians, in *Experimental Embryology of Marine and Fresh-water Invertebrates* (ed G. Reverberi), Amsterdam, North Holland, pp. 507–550.

SATOH, N. (1994) *Developmental Biology of Ascidians*, Cambridge University Press, New York.

VAN BENEDEN, E., JULIN, C.H. (1884) La segmentation chez les ascidens dans ses rapports avec l'organization de la larve, *Arch Biol*, **5**, 111–126.

WHITTAKER, J.R. (1973) Segregation during ascidian embryogenesis of egg cytoplasmic information for tissue-specific enzyme development. *Proc Natl Acad Sci USA*, **70**, 2096–2100.

CHAPTER 2

THE DEVELOPMENT
OF TADPOLE LARVAE
AND SESSILE JUVENILES

How does a multicellular, complex metazoan body generate from a single cell, the fertilized egg? Are there any common general rules for the formation of various body parts among chordates? We repeatedly emphasize that *Ciona* is one of the most appropriate systems for the study of chordate developmental genomics. In order to lay the groundwork for discussion in the following chapters, ascidian embryogenesis is briefly described here. A major reason that ascidians have long attracted embryologists is their relative embryonic simplicity as compared to vertebrates and other deuterostomes, such as sea urchins. At the same time, ascidian embryos are sufficiently complex to enable challenging developmental biology experiments as described in this book.

2.1 EMBRYOGENESIS AND TADPOLE LARVAE

2.1.1 Fertilization, Ooplasmic Movements, and Embryonic Axis Formation

Ciona intestinalis and *Halocynthia roretzi* eggs are 130–150 μm and 280 μm in diameter, respectively. The eggs are enclosed by a noncellular vitelline coat or chorion. On the outer surface of the vitelline coat are attached follicle cells. Within the perivitelline space between the egg and the vitelline coat there are many test cells. The existence of test cells within the space is a unique phenomenon in the animal kingdom. Test cells are likely to be involved in oogenesis and in the formation of the larval tunic.

The animal embryo is characterized by three body axes; anterior–posterior (A–P; also termed cranial–caudal), dorsal–ventral (D–V), and left–right (L–R) axis. However, the axis first visible in the ascidian egg is the animal–vegetal (An–Vg) axis. The animal pole is defined by the position in the egg where the polar bodies form, and the vegetal pole is defined as the position opposite the animal pole. Although the polar bodies become visible after fertilization, the An–Vg axis is established during oogenesis.

In ascidians, fertilization evokes two phases of "ooplasmic segregation or movement," whereby the egg cytoplasm undergoes dynamic rearrangement. During the first phase, mitochondria, myoplasm, and postplasmic/posterior end mark (PEM) messenger RNAs become concentrated at the vegetal pole in a microfilament-dependent manner. During

—————————
Developmental Genomics of Ascidians, First Edition. Noriyuki Satoh.
© 2014 John Wiley & Sons, Inc. Published 2014 by John Wiley & Sons, Inc.

the second phase, the vegetally localized cytoplasm is translocated to the future posterior pole, in a region called the posterior vegetal cytoplasm/cortex (PVC). Thus, eggs show transition from radial to bilateral symmetry. This second reorganization is driven by microtubules emanating from the sperm-derived centrosome. It has been shown that the location where the sperm first contacts the egg surface dictates the site of eventual penetration and position of the PVC. Ooplasmic movement is described in more detail in Chapter 5 of this book.

2.1.2 Cleavage

The cleavage of ascidian eggs is invariant and bilaterally symmetrical. The first plane divides the right and left halves. The first and second cleavages are perpendicular, dividing the egg into two and four cells, respectively, usually of equal sizes (Fig. 1.1c−e and Fig. 2.1). The third cleavage is latitudinal or horizontal, and usually divides cells slightly unequally. The division planes tend to slant, resulting in an H-shaped pattern when observed from the poles (Fig. 2.1). The four upper cells lie slightly anterior to the four lower cells. Thus, from the 8-cell stage onward, the A–P, D–V, and L–R axes become obvious. At the fourth cleavage the division pattern of the animal-hemisphere tier of four cells is reversed, as compared with that of the vegetal tier of four cells. All eight cells of the animal hemisphere are of nearly equal sizes, whereas those of the vegetal hemisphere develop to different sizes (Fig. 2.1 and Fig. 2.2b). Thereafter, the blastomeres divide under invariant patterns (Fig. 2.1). The timing of cell division becomes asynchronous after the 16-cell stage, temporarily resulting in 44-cell and 76-cell embryos (Fig. 2.1). From the fourth to the seventh cleavage, the posterior-most vegetal cells divide unequally, producing smaller daughter cells positioned posteriorly (Fig. 2.1 and Fig. 2.2b−e). Divisions of the animal blastomeres are synchronous, reflecting the clonal organization of ectodermal cells. By contrast, the vegetal blastomeres have different temporal division patterns, and these cells give rise to various tissues of mesodermal and endodermal origin.

2.1.3 Embryonic Cells Are Distinguishable and Named

Before gastrulation, which takes place at approximately the 112-cell stage, each blastomere has an outer surface (Fig. 1.1c−g, Fig. 2.1, and Fig. 2.2a−e). With a characteristic shape and cytoplasmic color pattern, each blastomere of the early embryo is easily distinguishable, and is named according to Conklin's scheme (1905), for example, a4.2, b5.3, A6.1, and B7.4 (Fig. 2.2). The letters "a" and "b" denote descendants of the two anterior and posterior animal blastomeres of the 8-cell embryo, respectively, while "A" and "B" indicate descendants of the anterior and posterior vegetal blastomeres, respectively (Fig. 2.2a). The first numerical digit designates the cell generation, counting the unsegmented egg as the first. The second digit annotates the cell with its own number, which doubles at each division (e.g., B7.4 divides into B8.7 and B8.8). Cells that lie nearer the vegetal pole are assigned the lower number. The blastomeres on the right side of the bilaterally symmetrical embryo are indicated by an underline. The uncleaved zygote is named A1. Blastomeres of the 2-cell embryo are termed AB2 (left) and <u>AB2</u> (right). The anterior and posterior blastomeres of the 4-cell embryo are named A3 and B3, respectively.

For example, B4.1 of the 8-cell embryo divides unequally to produce a smaller daughter, B5.2, at the posterior-most position of the 16-cell embryo (Fig. 2.2b). In turn, B5.2

produces the smaller cell, B6.3 (Fig. 2.2e), which then yields an even smaller cell posteriorly, B7.6. In *Ciona*, B7.6 gives rise to B8.11 and B8.12 during gastrulation; the latter cells undergo two more divisions at the late tailbud stage. B8.12 descendants are germ cell precursors (see Chapter 14). Conklin's nomenclature has now been extended to later stage divisions, allowing detailed analyses of cell-specification mechanisms. For example, A13.473 and A13.474 are neuronal cells of the visceral (motor) ganglion in the late tailbud embryo that possess different regulatory gene activities (see Chapter 10).

2.1.4   Gastrulation, Neurulation, and Tailbud Embryo Formation

Gastrulation occurs around the 112-cell stage and is followed by neurulation and tailbud embryo formation (Fig. 1.1h–l and Fig. 2.1). Early ascidian embryos have no conventional blastocoel so that animal and vegetal blastomeres contact each other inside the embryo after cell divisions. During gastrulation, endodermal cells and mesodermal cells in the vegetal hemisphere ingresses into the interior as the ectodermal layer migrates toward the vegetal pole to envelope the embryo. The spatial relationships among cells derived from the vegetal hemisphere are basically retained during these morphogenetic cell movements.

Neural plate formation initiates before the completion of blastopore closure. Because neurulation takes place at the vegetal side, the left–right assignment of the ascidian embryo is inverted relative to that of vertebrate embryos when looked down from the animal pole. The neural plate consists of distinctly arranged rows of cells (see Chapter 10). As in vertebrates, ascidian neurulation is accomplished by curling up the neural plate dorsally to form the neural tube (Fig. 1.1i, j and Fig. 2.1). The formation of this tube-like structure progresses from posterior to anterior. Once it is completely closed, the tail becomes elongated.

Upon tailbud formation (Fig. 2.1), the embryo enters the relatively long-lasting tailbud stage, during which the tail continues to elongate until the embryo is ready to hatch. Inside the embryo, the notochord cells converge at the dorsal midline. After the notochord cells are intercalated and stacked in a single row, they further extend individually along the A–P axis, providing a driving force for tail elongation. The cellular and molecular mechanisms of convergence and extension are conserved between ascidians and vertebrates.

2.1.5   Larval Morphology

The tadpole larva is composed of the trunk and tail (Fig. 1.1m, n and Fig. 2.2f–h). The overall anatomy of the ascidian larva is similar to that of its amphibian counterpart but comprises only ~2600 cells. This small number of constituent cells allows us to investigate the molecular mechanisms underlying cell differentiation and morphogenesis at the single-cell level.

The outer surface is formed from a layer of ~620 epidermal cells (Fig. 2.2f). Three papillae or palps represent adhesive organs at the anterior tip of the trunk. The dorsal part of the trunk is composed of the central nervous system (CNS), with approximately 170 neurons and 230 ependymal (glial) cells (Fig. 2.2f). The CNS is subdivided into the sensory vesicle, posterior brain, neck region, visceral ganglion, and nerve cord. The nerve cord consists of only four rows (one dorsal, two lateral, and one ventral) of ependymal cells (Fig. 2.2h), ~8 neurons and bundles of axons, and extends to the posterior part of the tail. The sensory vesicle has two pigment cells (Fig. 1.1m, n).

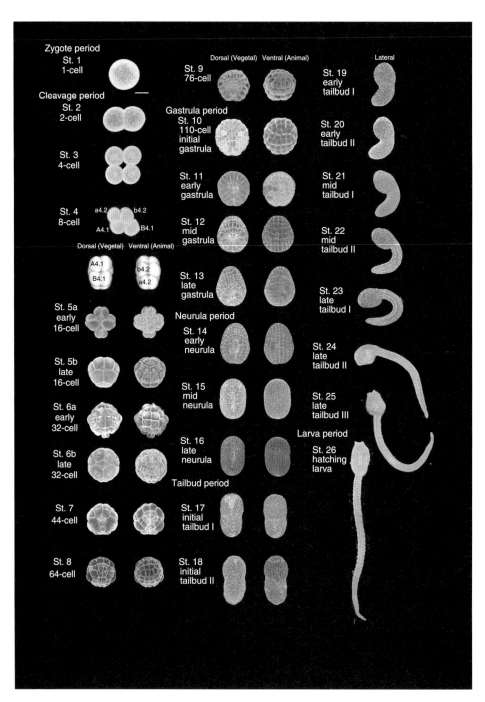

One is a gravity-sensory organ, the otolith, and the other is a component of the light-sensory organ, the ocellus. These two are distinguished by different pigmentation patterns. The peripheral nervous system constitutes 20 to 30 epidermal sensory neurons that are distributed on the midline of the larval body surface.

The ventral part of the larval trunk is occupied by endodermal cells, and endodermal tissue extends to the tail region as the endodermal strand, which includes primordial germ cells in the posterior region (Fig. 1.1n and Fig. 2.2f). The mesenchymal cells are situated bilaterally in the posterior region of the trunk. The trunk lateral cells (TLCs) and trunk ventral cells (TVCs) are small groups of mesodermal cells located bilaterally to the dorsal and ventral sides, respectively (Fig. 2.2g). TLCs contribute to adult mesodermal organs and tissues, while TVCs are precursors of the juvenile heart.

In the center of the tail, the notochord, which consists of exactly 40 cells, runs from anterior to posterior (Fig. 1.1m, n and Fig. 2.2f), providing a supportive rod when the flanking muscle contracts and relaxes for tail movement. The muscle cells are aligned in three rows on each side of the notochord (18 and 21 cells on each side in *Ciona* and *Halocynthia* embryos, respectively) (Fig. 1.1n and Fig. 2.2g) and generate waves of bilateral motion in the tail. The muscle cells are striated but not fused like those in vertebrates.

Despite such detailed description of larval constituent cells, there are still several cell types, whose function remains to be defined in future.

2.1.6 Cell Lineage and Restriction of Developmental Fates

An advantage of the ascidian embryo for developmental biology studies is that the cleavage profiles are invariant between individuals, and embryonic cell lineages are well traced and documented (Fig. 2.3). Another prominent feature of ascidian embryogenesis is that the developmental fates of embryonic cells are restricted at very early stages (Fig. 2.2). As early as the 16-cell stage, a pair of epidermis-restricted cells (b5.4 pair) appears in the animal hemisphere (Fig. 2.2b). Then, at the 32-cell stage, two pairs of cells in the vegetal hemisphere (A6.1 and B6.1 pairs) become restricted to endoderm (Fig. 2.2c). By the 64-cell stage, blastomeres appear that are restricted to their various individual fates: notochord, muscle, mesenchyme, and TLCs (Fig. 2.2d). And by the 112-cell stage, fate

FIGURE 2.1 Staging of *Ciona intestinalis* embryogenesis. Three-dimensional reconstructed images of the embryo in a postfertilization developmental time course. The top of each embryo is anterior, except for the embryos from Stage 24 where the left side is anterior. Stage 1 in the zygote period. Stages 2–9 in the cleavage period (Stage 2, 2-cell embryo; Stage 3, 4-cell embryo; Stage 4, 8-cell embryo; Stage 5, 16-cell embryo; Stage 6, 32-cell embryo; Stage 7, 44-cell embryo; Stage 8, 64-cell embryo; and Stage 9, 76-cell embryo). Stages 10–13 in the gastrula period (Stage 10, initial gastrula; Stage 11, early gastrula; Stage 12, mid-gastrula; and Stage 13, late gastrula). Stages 14–16 in the neurula period (Stage 14, early neurula; Stage 15, mid-neurula; and Stage 16, late neurula). Stages 17–25 in the tailbud period (Stage 17, initial tailbud I; Stage 18, initial tailbud II. The tail begins to bend; Stage 19, early tailbud I; Stage 20, early tailbud II; Stage 21, mid-tailbud I. Stage 22, mid-tailbud II; Stage 23, late tailbud I—11.9 h; Stage 24, late tailbud II. The otolith pigmentation is initiated and palps formation starts; and Stage 25, late tailbud III. Ocellus melanization occurs). Stage 26 in the larva period. Scale bar 50 μm. For further details see http://chordate.bpni.bio.keio.ac.jp/faba/top.html. (Adapted from Hotta *et al.* (2007) *Dev Dyn*, **236**, 1790–1805.)

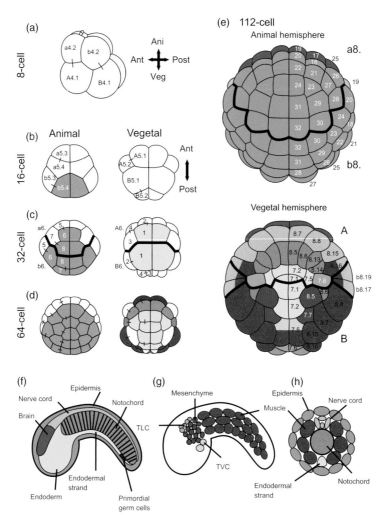

FIGURE 2.2 Cleavage, cell lineage, and gradual restriction of developmental fates of ascidian embryonic cells. Blastomeres are named according to Conklin's nomenclature and colored when the developmental fate is restricted to a single type of tissue. (a) An 8-cell embryo, lateral view. The animal pole is up and the vegetal pole is down. Anterior is to the left and posterior is to the right. Bars indicate sister-cell relationship. (b) A 16-cell embryo, viewed from the animal (left) and vegetal poles (right). Anterior is up and posterior is down. (c) A 32-cell embryo, animal (left) and vegetal (right) views. (d) A 64-cell embryo, animal (left) and vegetal (right) views. (e) A 112-cell embryo, animal (right) and vegetal (left) views. Fate restriction is shown in the animal hemisphere (top) and vegetal hemisphere (bottom). Bold lines indicate the boundaries between the a-, b-, A-, and B-line cells. (f–h) Schematic drawing showing tissues and organs of the tailbud embryo. (f) Midsagittal section and (g) sagittal section of the embryo, and (h) transverse tail section. TLCs, trunk lateral cells. TVCs, trunk ventral cells. (Modified from Nishida (2005) *Dev Dyn*, **233**, 1177–1193.)

restriction is accomplished in 102 blastomeres (Fig. 2.2e). For example, the following cells have two or more fates at the 112-cell stage: a8.19 (brain/primordial pharynx), b8.17 (muscle/nerve cord/endodermal strand), b8.19 (muscle/nerve cord), A8.18 (nerve cord/muscle), and B7.5 (muscle/TVC). The further restriction of these fates as embryogenesis proceeds has been mapped. There is effectively a blue print for the chordate body-plan outlining the origins of every embryonic cell.

2.1.6.1 Fates of a-Line Cells

The a-line cells, which are descendants of the a4.2 quadrant of the 8-cell embryo, mainly give rise to the larval trunk epidermis and to the nervous system, including the brain. As shown in Fig. 2.3, the developmental fates of a6.6, a6.8, a7.14, and a8.26 are restricted to the epidermis. Both a6.6 and a6.8 give rise to 64 epidermal cells each. a8.18 and a8.20 develop into the left dorsal palp and the left half of the ventral palp, respectively. a8.17 produces the anterior and posterior parts of the brain, while a8.19 generates the primordial pharynx and the ventral part of the brain. a8.25 gives rise to pigment cells of the sensory organs, the dorsal and lateral parts of the

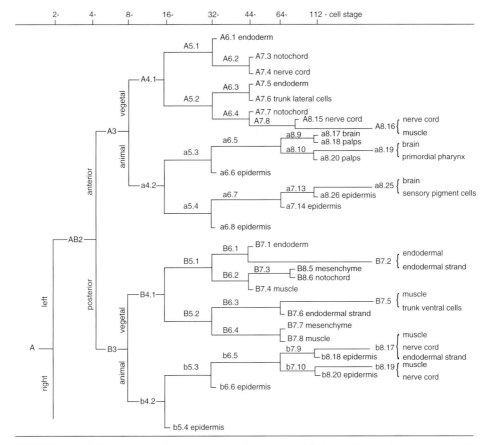

FIGURE 2.3 Cell lineages and developmental fate segregation in ascidian embryos. Because the lineage is bilaterally symmetrical, only the left half of the embryo is shown. When the developmental fate of a certain blastomere becomes restricted to one type of tissue, further divisions of that blastomere are abbreviated. Further fate restriction will be discussed in related chapters.

brain, and several undefined cells. The bilateral pair a8.25 and <u>a8.25</u> yield the otolith and ocellus in a complementary manner. The details of the CNS lineage, and the molecular mechanisms underlying its cell fate specification are discussed in Chapter 10.

2.1.6.2   Fates of b-Line Cells   The b-line cells, descendents of the b4.2 quadrant of the 8-cell embryo, mainly give rise to the larval tail epidermis (Fig. 2.3). The developmental fates of b5.4, b6.6, b8.18, and b8.20 are restricted to the epidermis, and they form 128, 64, 16, and 16 epidermal cells, respectively. b8.17 descendents develop into the muscle, nerve cord, and endodermal strand in the caudal tip. b8.19 yields muscle and nerve cord in the tail of *Halocynthia* larvae, but it does not give rise to any muscle in *Ciona* larvae.

2.1.6.3   Fates of A-Line Cells   The A-line cells are descendants of the A4.1 quadrant. As shown in Fig. 2.3, A6.1, and A7.5 are restricted to produce endodermal cells. A7.3 and A7.7 are destined to become notochord, and each produces eight notochord cells after three divisions. A7.4 develops into the ventral row of the nerve cord, and A7.6 generates TLCs. A8.15 gives rise to the lateral row of nerve cord cells, and A8.16 progeny become muscle and nerve cord cells at the tail tip (Fig. 2.3).

2.1.6.4   Fates of B-Line Cells   The B-line cells, descendants of the B4.1 quadrant, give rise to various endomesodermal organs (Fig. 2.3). B7.1 is a precursor of endoderm in the trunk of the tailbud embryo, while B7.2 gives rise to ventral endoderm and all but the caudal tip of the endodermal strand. B7.6 is a primordial germ cell precursor. B7.4 and B7.8 yield muscle cells. B7.7 and B8.5 engender mesenchymal cells. B7.5 gives rise to muscle and TVCs, the latter of which are adult heart muscle progenitors.

2.1.7   Developmental Stages of Embryogenesis

To facilitate a more precise discussion of *Ciona* embryogenesis, a developmental staging has been defined based on detailed time-lapse recording and 3D image reconstructions of fixed embryos (Fig. 2.1; see http://chordate.bpni.bio.keio.ac.jp/faba/top.html). This series includes 26 distinct stages from Stage 0 (unfertilized eggs) to Stage 26 (hatching larvae). Stages 2 to 9 cover the cleavage period, Stages 10 to 13 the gastrula period, Stages 14 to 16 the neurula period, Stages 17 to 25 the tailbud embryo period, and Stage 26 the larval period (Fig. 2.1). Together with further 3D images of the *Ciona* tailbud embryos, this table also provides the configuration of every constituent embryonic cell at each stage.

2.2   METAMORPHOSIS AND MAKING A SESSILE JUVENILE

2.2.1   Metamorphosis

The metamorphosis of tadpole-like swimming larvae to sessile juveniles and adults involves dynamic and complex changes in the shape and physiological function of constituent cells. The preoral lobe, including the papilla, plays a role in metamorphosis initiation in response to a wide variety of external and endogenous signals. Metamorphosis requires several major developmental events including (i) adhesion of larvae to substratum by papillae, (ii) absorption of the tail, (iii) loss of the outer cuticle layer of the larval tunic, (iv) emigration of blood cells or pigmented cells, (v) rotation of visceral organs through an arc of about $90°$ and expansion of the branchial basket, (vi) expansion, elongation, or reciprocation of ampullae, reorientation of test vesicles, and expansion of the tunic, (vii) retraction of the sensory vesicle, and (viii) release of organ rudiments from an arrested state of development.

The process of *C. intestinalis* juvenile formation is characterized by nine stages starting with swimming larvae (Stage 0) and proceeding through juvenile development to the second ascidian stage (completion of metamorphosis) (Stage 8). The staging is based on the development of the oral siphon, tentacle, oral and atrial pigments, atrial siphon, ganglion and neural gland, longitudinal muscle, stigmata, transverse bar and languet, longitudinal bar and papilla, heart, digestive organ, gonad, endostyle, stalk, and villi. These descriptions and the staging system provide a basis for studying the cellular and molecular mechanisms underlying the development of adult organs and tissues in this basal chordate.

2.2.2 Larval to Juvenile Organs

The larval organs can be categorized for didactic purposes into two major groups: transitory larval organs (TLOs) and prospective juvenile organs (PJOs). TLOs function in larval locomotion (notochord and muscle), sensory input (otolith and ocellus), and settlement (papillae). They are fully differentiated in the larva, but are destroyed or, in the case of the fins, lost during metamorphosis. By contrast, PJOs may be in an arrested state of development in the larva and become functional either shortly after settlement or following further histogenesis.

The development of PJOs has been analyzed by recent larval cell tracing studies. For example, larval trunk epidermal cells give rise to adult epidermis. Similarly, endodermal cells give rise to various juvenile endodermal organs, including the endostyle, branchial sac, digestive organs, peripharyngeal band, and dorsal tubercle (see Chapter 7).

The boundaries between clones descended from early blastomeres do not correspond to the boundaries between adult endodermal organs. However, the topographic position of each prospective region in the larval endoderm is similar to that of the adult organs, indicating that marked rearrangement of endodermal cell positions does not occur during metamorphosis (see Chapter 7). On the other hand, mesodermal juvenile tissues come from mesodermal larval tissues such as mesenchyme, TVCs, and TLCs (see Chapter 11). In spite of the terminology, larval mesenchymal cells do not contribute to the formation of mesodermal juvenile organs. Instead, they yield juvenile tunic cells in *Halocynthia* and give rise to tunic and blood cells in *Ciona*. TVCs generate body wall muscle (atrial siphon and latitudinal mantle), heart, and juvenile pericardium (see Chapter 13). In *Halocynthia*, TLCs give rise to blood (coelomic) cells, oral siphon and longitudinal mantle muscle, and epithelium of the first and second gill slits. In *Ciona*, TLCs give rise to a part of the stomach, but not to longitudinal mantle muscle. Therefore, although differences exist in the fates of larval mesenchymal cells and TLCs between *Halocynthia* and *Ciona*, the developmental fates after metamorphosis are almost invariant with respect to the mesodermal organs and progeny cell types.

2.2.3 Adult Morphology

The adult morphology was explained in Chapter 1. The following provides additional information regarding adult organs and/or tissues. An individual has two openings: an incurrent branchial (oral) siphon and an outcurrent atrial siphon (Fig. 1.1a, b). Each siphon can be closed by sphincter muscles and has ocelli along its rim. The mouth behind the branchial siphon leads to a large pharynx or branchial basket, a chamber perforated by dorsoventral rows of numerous and ciliated gill slits called stigmata (Fig. 1.1b). Blood vessels traverse the pharyngeal wall between the slits.

Along the ventral margin of the branchial basket is a specialized organ called the endostyle that secretes large quantities of mucus used for capturing food particles (Fig. 1.1b). The endostyle contains iodine, and therefore this organ has an evolutionary relationship to the vertebrate thyroid gland. The digestive tract leads to a stomach, followed by an intestine that terminates at the anus, which opens into the atrial cavity (Fig. 1.1b). The nervous system contains a single cerebral ganglion lying between the two siphons (Fig. 1.1b), from which several nerves elongate to various parts of the body, for example, the muscles, pharynx, viscera, reproductive organ, and siphons. A neural or subneural gland adjacent to this ganglion leads through a duct to the pharynx, just behind the mouth. The neural gland may be considered as the forerunner of the vertebrate pituitary gland. The open circulatory system is well developed and consists of a short, tubular heart and numerous blood vessels without endothelial cells. Several different types of blood cells or coelomic cells are distinguished, including hemocytes, lymphocytes, amoebocytes, signet-ring cells, vacuolated cells, and pigment cells. Ascidian blood has several specialized functions, such as the accumulation of vanadium. Recent studies show that signet-ring cells are involved in vanadium accumulation.

SELECTED REFERENCES

CHIBA, S.., SASAKI, A., NAKAYAMA, A., TAKAMURA, K. *et al.* (2004) Development of *Ciona intestinalis* juveniles (through 2nd ascidian stage). *Zoolog Sci*, **21**, 285–298.

CLONEY, R.A. (1990) Urochordata-Ascidiacea, in *Reproductive Biology of Invertebrates* (eds K.G. Adiyodi and R.G. Adiyodi), Oxford and IBH, New Delhi, pp. 361–451.

CONKLIN, E.G. (1905) The organization and cell lineage of the ascidian egg. *J Acad Natl Sci (Philadelphia)*, **13**, 1–119.

HOTTA, K., MITSUHARA, K., TAKAHASHI, H., INABA, K. *et al.* (2007) A web-based interactive developmental table for the ascidian *Ciona intestinalis*, including 3D real-image embryo reconstructions: I. From fertilized egg to hatching larva. *Dev Dyn*, **236**, 1790–1805.

NAKAMURA, M.J., TERAI, J., OKUBO, R., HOTTA, K. *et al.* (2012) Three-dimensional anatomy of the *Ciona intestinalis* tailbud embryo at single-cell resolution. *Dev Biol*, **372**, 274–284.

NISHIDA, H. (1987) Cell lineage analysis in ascidian embryos by intracellular injection of a tracer enzyme. III. Up to the tissue restricted stage. *Dev Biol*, **121**, 526–541.

NISHIDA, H. (2005) Specification of embryonic axis and mosaic development in ascidians. *Dev Dyn*, **233**, 1177–1193.

SATOH, N. (2013) *Ciona*: a model for developmental genomics, in *Encyclopedia of Life Sciences*. John Wiley & Sons, Ltd, Chichester. doi: 10.1002/9780470015902.a0021411.

CHAPTER 3

GENOMICS, TRANSCRIPTOMICS, AND PROTEOMICS

In today's bioinformatics-driven research environment, the power of *Ciona* as a model organism for developmental genomics is underscored. Researchers can obtain datasets that describe the temporal and spatial expression of developmentally relevant genes, and use those as a starting point for determining the molecular mechanisms that underlie phenomena of interest. The goal is to discover the "common words" across organismal complexity that are responsible for the establishment of the metazoan body plan. To what extent we can tackle this challenge depends largely on the strength of the relevant infrastructure. This chapter introduces ascidian genomics, transcriptomics, and proteomics, which have thus far been characterized mainly in *Ciona intestinalis*. As a model system, *Ciona* is a standout that contains plentiful, high quality datasets. Several websites related to ascidian bioinformatic resources are supplied.

3.1 THE GENOME OF *Ciona intestinalis*

3.1.1 General Features

The draft genome of *C. intestinalis* was published in 2002, which at that time it was the seventh animal genome to be sequenced. A single individual was used as the DNA source for sequencing. The ~159 Mb genome is composed of ~117 Mb of nonrepetitive and euchromatic sequence, ~18 Mb of high copy tandem repeats (rRNA, tRNA clusters, and others), and ~17 Mb of additional low copy transposable elements and repeats. The genome is notably AT rich (65%) compared with the human genome (45%). A high level of allelic polymorphisms, with 1.2% of the nucleotides differing between alleles, was observed.

Several genome models have been predicted and published at http://ghost.zool.kyoto-u.ac.jp/cgi-bin/gb2/gbrowse/kh/ (KH gene model), http://www.jgi.doe.gov/ciona, and http://www.ensemble.org. As discussed later, as of January 1, 2013, a total of 1,205,674 expressed sequence tags (ESTs) have been registered in the National Center for Biotechnology Information (NCBI) database of expressed sequence tag (dbEST) database (http://www.ncbi.nlm.nih.gov/dbEST/dbEST_summary.html). Although this species ranks 12th in terms of the number of public entries, the top ten organisms, including *Homo sapiens*, *Mus musculus*, and *Zea mays* (maize), have comparatively larger genomes with more transcripts than *Ciona*. The KH gene model set that is based

Developmental Genomics of Ascidians, First Edition. Noriyuki Satoh.
© 2014 John Wiley & Sons, Inc. Published 2014 by John Wiley & Sons, Inc.

TABLE 3.1 **The Number of Genes and Transcripts in *Ciona intestinalis* by the KH Gene Model**

Predicted gene loci	15,254
Predicted transcripts	24,025
Transcripts that putatively encode the full ORF[a]	20,239
Transcript 5′-ends identified by SL[b] ESTs	11,797
Transcript 5′-ends identified by nonSL oligocapping ESTs	818
In-frame stop codons in the 5′-region of the longest ORFs of transcripts not represented by 5′-full-length ESTs	7,624
Operons	1,310
Operon genes	2,909

[a] Open reading frame.

[b] Spliced leader.

on a total of 1,391,262 ESTs and cDNAs (1,179,850 conventional ESTs + 202,535 full-length ESTs + 8877 full-insert cDNA sequences) is widely accepted as the best prediction of *Ciona* genes. According to the model, the *Ciona* genome contains 15,254 loci and 24,025 transcripts (Table 3.1).

C. intestinalis genes have at least three characteristic features. First, the *Ciona* genome contains a basic set of chordate genes including those for developmentally important TFs and SPMs or signal transduction molecules (STMs). It has been shown that genome-wide gene duplication occurred twice during the evolution of the vertebrate lineage. Although many of the duplicates have since disappeared, most of those in the TF and SPM families are likely to have been retained. The duplication of these genes may be a major reason why vertebrates evolved more complex body plans. However, duplicated genes frequently show overlapping expression and functional redundancy, hindering functional investigations. By contrast, the *Ciona* genome contains the basic ancestral chordate complement of genes that were in place prior to the second round of gene duplication (Table 3.1). For example, as shown in Fig. 3.1, vertebrates contain five typical mitogen-activated protein kinase (MAPK) subfamilies, including ERK1/ERK2, JNK, p38, ERK3/ERK4, and ERK5, as well as one slightly diverged MAPK subfamily, Nemo-like kinase (NLK). A phylogenetic analysis using a kinase domain reveals six genes in the *Ciona* genome, which are confidently predicted to represent each of the subfamily. *Ci-JNK* was a *Ciona* ortholog of *JNK* (bootstrap value of 100%), *Ci-p38* for *p38* (100%), *Ci-ERK1/2* for *ERK1/ERK2* (100%), *Ci-ERK3/4* for *ERK3/ERK4* (100%), *Ci-ERK5* for *ERK5* (100%), and *Ci-NLK* for NLK (100%) (Fig. 3.1). Thus, the *Ciona* genome contains a complete set of MAPKs found in vertebrates with no redundancy. The expression of these genes was evidenced by corresponding ESTs. This relative simplicity makes the *Ciona* system ideal for testing gene expression and function, as is discussed in Chapter 4.

Second, there are several evolutionarily lost genes in the *Ciona* genome. The *Hox* genes of *C. intestinalis* are not located within a single cluster but rather have been split onto chromosomes 1 and 7 (Fig. 3.3). In addition, *Hox7*, *8*, and *9* appear to have been lost during retrograde evolution of the ascidians. Third, there are several genes that are specific to the ascidian lineage. As previously mentioned, urochordates are the only animals that can synthesize cellulose independently. The *Ciona* genome contains a gene for cellulose synthase, which was likely obtained via horizontal transfer from bacteria (see Chapter 8).

MAPK

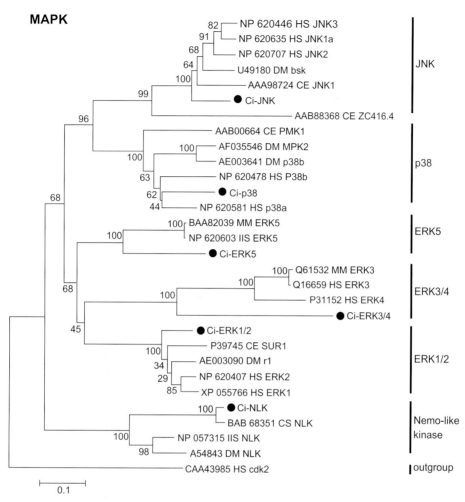

FIGURE 3.1 Phylogenetic tree of MAPK generated by the neighbor-joining method based on the alignment of a kinase domain. *Ciona intestinalis* proteins are shown by large black dots. The number matching each branch indicates the percentage of times that a node was supported in 1000 bootstrap pseudoreplications. Human CDK2 kinase was used as an outgroup protein. The scale bar indicates an evolutionary distance of 0.1 amino acid substitutions per position. (Adapted from Satou *et al.* (2003a) *Dev Genes Evol*, **213**, 254–263.)

3.1.2 The *Ciona* Genome Is Compact

The presence of ~15,250 protein-coding genes in the 117-Mb euchromatic genome means that one gene occupies 7.7 kb on average. Accordingly, *Ciona* genes are organized in the genome more compactly than those of the protostomes (except *Caenorhabditis elegans*), deuterostomes, and vertebrates, indicating an advantage of the *Ciona* system for future studies of *cis*-regulatory networks. The average gene density of *Ciona* (one gene/7.7 kb) is slightly more compact than *Drosophila* (one gene/8.9 kb), and much more so than mice (one gene/113 kb).

In addition, the genome of *Ciona savignyi*, a closely related species of *C. intestinalis*, has also been sequenced. Its size is estimated to be ~190 Mb (http://www.broad .mit.edu/annotation/ciona/), and the sequenced individual had an extremely high heterozygosity rate, averaging 4.6% with significant regional variation and rearrangements at all physical scales (http://mendel.stanford.edu/sidowlab/ciona.html). Together with this heterozygosity, the high quality genomic sequences of the two *Ciona* species allow for easy detection of conserved noncoding sequences, which are likely to be involved in the regulation of gene expression (see Chapter 4). Genome-decoding projects are now going on in other ascidians including *Halocynthia roretzi* and *Phallusia mammillata*, which provide promising comparative genomics of ascidians.

3.1.3 The *Ciona* Developmentally Relevant Genes Are Well-Characterized

The *C. intestinalis* genome is among the most comprehensively annotated with respect to developmentally relevant genes, for which the information is substantiated by a considerable amount of cDNA information (Table 3.2 and Table 3.3). The genome contains ~637 TF genes. Of them, 318 genes display well-conserved motifs (Table 3.2), including 47 genes for basic helix-loop-helix (bHLH) family members, 26 for basic leucine zipper domain (bZIP) family members, 15 for E-twenty six (Ets) family members, 29 for forkhead box (Fox) family members, 21 for high mobility group (HMG) family members, 93 for homeobox family members, 18 for nuclear receptor family members, 10 for

TABLE 3.2   The Number of Transcription Factor Genes in *Ciona intestinalis*

Family	Number of genes annotated	Number of genes mapped on chromosomes	Number of genes with analyzed expression profiles
bHLH	47	33	40
bZIP	26	18	24
Ets	15	14	15
Fox	29	20	24
HMG	21	15	20
Homeobox	93	70	74
NR	18	15	17
T-box	10	8	9
Other TF[a]	59	40	55
subtotal	318	233	278
Other TF[b]	21	14	21
subtotal	339	247	299
TF with zinc-finger motif[c]	298	181	250
Total	637	428	549

[a] Genes that encode sequence-specific TFs.

[b] Genes that may or may not encode sequence-specific TFs.

[c] This excludes TF genes that are grouped with other families such as NR.

TABLE 3.3   The Number of Major Genes Encoding Signaling Pathway Molecules in *Ciona intestinalis*

Family	Number of genes annotated	Number of genes mapped on chromosomes	Number of genes with analyzed expression profiles
RTK pathway			
Ephrin	5	5	4
Eph Receptors	5	5	5
Fibroblast growth factor (FGF)	7	4	6
FGF Receptor	1	0	1
Transforming growth factor (TGF)-β pathway			
Ligands	12	11	10
Type-I receptors	3	1	3
Type-II receptors	2	1	2
Wnt pathway			
Ligands	10	8	7
Frizzled receptors	5	4	3
Hedgehog pathway			
Ligands	2	2	2
Patched	1	1	1
Smoothened	1	0	2
Notch pathway			
Ligands	3	1	3
Notch receptor	1	1	1

NFkB and JAK-STAT are not listed here.

T-box family members, and 59 other sequence-specific TF genes (http://ghost.zool.kyoto-u.ac.jp/TF_KH.html). The *Ciona* genome also contains 21 genes with potential TF activity and 298 possible TF genes with zinc-finger motifs.

In addition, the *Ciona* genome contains genes encoding major SPMs (Table 3.3). Specifically, a total of 58 are well-annotated genes, including 18 for the receptor tyrosine kinase (RTK)/MAP kinase pathway, 17 for the transforming growth factor (TGF)-β pathway, 15 for Wnt signaling, four for hedgehog signaling, and four for the Notch cascade (http://ghost.zool.kyoto-u.ac.jp/TF_KH.html).

3.1.4   The Genomic Information Is Mapped onto Chromosomes

One of the most challenging questions in developmental genomics is how the body plan construction gene regulatory network (GRN) is controlled at the chromosomal level and/or genome-wide. *C. intestinalis* has 14 pairs of chromosomes, most of which are telocentric. Approximately 82% of the 117 Mb nonrepetitive sequences have been mapped onto all of the chromosomal arms by two-color fluorescent in situ hybridization (FISH) of 273 representative bacterial artificial chromosome (BAC) clones (Fig. 3.2). The mapped genes include 302 of the 318 core TF genes (95%; Fig. 3.3) and 111 of the 125 major SPM genes (89%; Fig. 3.3).

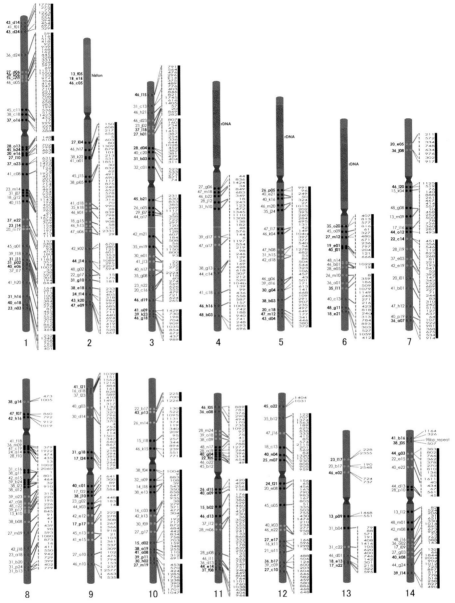

FIGURE 3.2 Mapping of BAC clones and genome sequences onto 14 pairs of *Ciona intestinalis* chromosomes. The length of the diagrammatic chromosome arms represents their approximate relative genome size. The location of 273 FISH-mapped BACs is shown relative to DAPI-counter-stained chromosomes with centromere bands. Clones that are mapped proximal to the centromere are represented by a pair of black circles, those mapped distal to the centromere are represented by a pair of red circles, and those mapped to the middle of the arms by a pair of gray circles. Red areas indicate the short arms of chromosomes 4, 5, and 6 with rDNA clusters, and the green area depicts the 5S rDNA and histone clusters. The paired-BAC end sequences assembled 807 scaffolds (96.6 Mb) into 46 joined-scaffolds, as shown by bold lines. Arrows on the left side of the scaffold number indicate the orientation of mapped sequences. (Adapted from Shoguchi *et al.* (2008) *Dev Biol*, **316**, 498–509.)

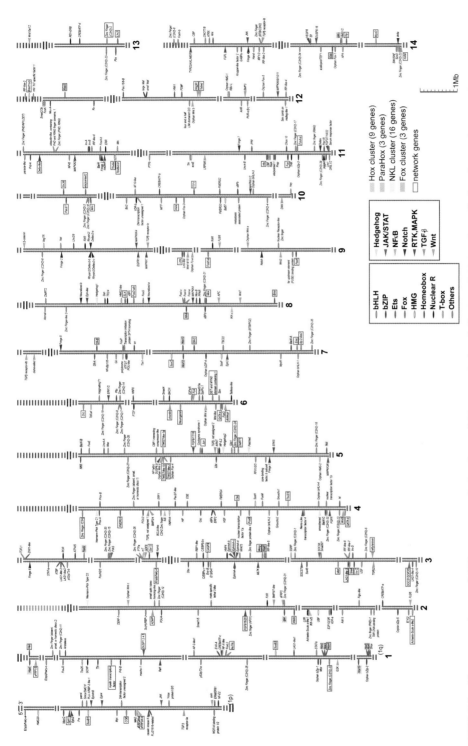

FIGURE 3.3   Chromosomal map of 373 core transcription factor (TF) genes (colored discs) and 111 major cell signaling pathway molecule (SPM) genes (colored arrowheads) in *Ciona intestinalis*. Centromeric regions are shown by dark blue dashed lines. Red dashed lines indicate three rDNA clusters, green dashed lines highlight a histone cluster region. The enclosed genes in red lines are those analyzed as elements of the regulatory network for chordate body plan construction (see Fig. 12.6). (Adapted from Shoguchi *et al.* (2008) *Dev Biol*, **316**, 498–509.)

25

The short arm of chromosome 2 is highly polymorphic in length, and consists of a cluster of 5S rDNA genes and histone genes. Similarly, the short arms of chromosomes 4, 5, and 6 also exhibit high polymorphism with rDNA clusters (Fig. 3.2). The short and long arms of the other chromosomes contain a similar number of genes per length, although genes that arose by duplication of a single ancestor appear to be clustered in the genome. In general, the 302 core TF genes and 111 major SPM genes appear to be distributed rather evenly over the 14 chromosomes (Fig. 3.3). This is the case for each gene family. For example, the *Ciona* genome contains 93 homeobox genes. Of the 91 that are mapped, 18 are present on chromosome 1, three on chromosome 2, and 12 on chromosome 3. The remaining 58 genes are mapped on chromosomes 4 through 14, suggesting no special localization or clustering of homeobox genes in the *Ciona* genome. A similar positioning profile is seen for the 111 major SPM genes, which include coordinated expressed signaling components in common signal transduction pathways. For example, 15 Wnt pathway genes are distributed over seven chromosomes, each of them containing one to five genes (Fig. 3.3).

3.2 TRANSCRIPTOMICS

3.2.1 EST and cDNA Data Sets

As mentioned earlier, as of January 1, 2013, a total of 1,205,674 ESTs were registered in the NCBI EST database. The ESTs cover transcripts expressed at different developmental stages, for example, fertilized eggs, cleaving embryos, gastrulae/neurulae, tailbud embryos, tadpole larvae, testis, ovary, endostyle, neural complex, heart, and blood (coelomic) cells, and whole young adults. ESTs were categorized into 17,834 independent cDNA clones or uni-genes, which cover ~74% of the transcripts expressed in *C. intestinalis*. The representative cDNAs that were rearrayed in 47,384 well plates have been distributed as "*Ciona intestinalis* Gene Collection Release 1 (CiGCR1)" for free academic research use. Every clone is available through the National Bioresource Center, RIKEN, Japan (http://www.nbrp.jp). Another Gateway full-open reading frame (ORF)-clone collection is also available through ANISEED (see Chapter 12).

3.2.2 Application of the EST Database

The size of the EST database provides insight into the temporal and spatial regulation of gene expression. The cDNA libraries used for EST analyses were not amplified or normalized, so the appearance of cDNA clones (or EST counts) occurs in proportion to their abundance at the corresponding stage or location. For example, if the EST count of a given gene is scored only in fertilized eggs and very early embryos but not at other developmental stages, this gene is highly likely to be expressed maternally. Alternatively, if an EST count is scored only in the heart muscle, the gene may be specifically expressed there. Assessments of gene expression levels by EST count correspond well to similar characterizations by Northern blot analysis and/or whole-mount in situ hybridization (WISH). Thus, EST counts provide a convenient approach for the initial description of temporal and spatial expression of the genes.

3.2.3   Spatial Expression of Developmentally Relevant Genes

Ciona is the best developmental model system available in terms of spatial gene expression profiles. The spatial expression profiles of almost all of the core TF genes and major SPM genes have been determined (data are available at the Ghost website and the ANISEED website). This has greatly advanced studies to identify regulatory genes involved in particular developmental processes. For example, of 93 for homeobox family members, 74 genes each have been clarified with their spatial expression profiles. For another example, a bHLH gene, *Mesp*, is expressed specifically in B7.6 and plays a pivotal role in juvenile heart formation. The isolation of *FoxF* as a downstream gene of *Mesp* was inspired by its expression in B7.6 daughter cells (see Chapter 13).

In addition, approximately 1000 genes were randomly selected from those expressed at the five developmental stages—fertilized eggs, cleavage-stage embryos, tailbud embryos, tadpole larvae (Fig. 3.4), and young adults. Each gene has been localized with respect to its spatial expression. All of the cDNA and spatial expression data are accessible at the Ghost website.

3.2.4   Splice Leader and Operons

The updated assembly and gene model set permit new insight into global features of the *Ciona* genome, including the nature of the intron population. For most of the 113,879 introns in the KH gene model set, the best alignments are consistent with the expected presence of the canonical donor (GT) and acceptor (AG) site dinucleotides, introns being spliced by U2 spliceosomes. However, for 596 introns the best alignments are not consistent with the canonical GT–AG site but consistent with the use of the known noncanonical dinucleotides GC–AG (556 introns), and AT–AC (40 introns), the latter of which are spliced by U12 spliceosomes.

It has been shown that the *Ciona* genome contains 2909 candidate operons, which represents approximately one-fifth of the total number of genes in the genome (see Fig. 6.4a). Consistent with the hypothesis that polycistronic preRNAs derived from operons are resolved into monocistronic mRNAs by splice leader (SL) trans-splicing, a very high proportion (1158 out of 1599, or 72%) of operon downstream genes are represented by 5′-full-length ESTs. 11,797 of 20,239 transcripts transcribed from the *Ciona* genome are modified by adding SL sequences (Table 3.1).

3.3   PROTEOMICS

Gene functions are generally realized via their protein products, but the mRNA and protein expression levels of a given gene are not always parallel. Therefore, characterizing a protein's localization, interactions, and function(s) is essential to understanding the molecular mechanisms underlying developmental processes. *Ciona* proteomic datasets, which are greatly supported by the available genomics and transcriptomics information, have recently appeared as CIPRO database (http://www.cipro.ibio.jp) (Fig. 3.5).

The CIPRO database is an integrated protein database for *C. intestinalis*. The current database contains 70,493 unique sequences covering all of the proteins from all of the genetic models for this species. The sequences of more than 5000 proteins have been manually annotated based on the large-scale transcriptomic, proteomic, and bioinformatic data. CIPRO also has original experimental data including 2D-PAGE gels with

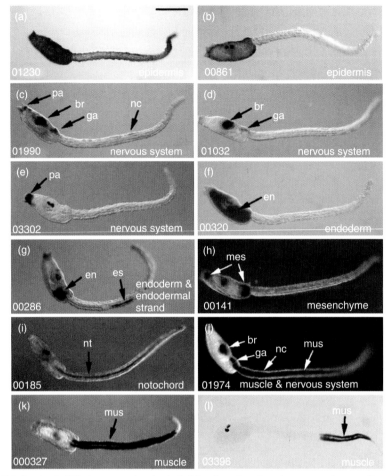

FIGURE 3.4 (a–l) Whole-mount in situ hybridization showing the expression profiles of genes specific to organs and/or tissues in *Ciona intestinalis* larvae. The organ where the gene is specifically expressed is shown in the right bottom corner, and the clone names in the left bottom corner. br, brain; en, endoderm; es, endodermal strand; ga, visceral ganglion; mes, mesenchyme; mus, muscle; nc, nerve cord; nt, notochord; pa, papillae. Scale bar represents 100 μm and is applicable to all photographs. (Adapted from Kusakabe *et al.* (2002) *Dev Biol*, **242**, 188–203.)

proteins identified by protein mass fingerprint (PMF) mass spectrometric (MS) analysis, and localization studies describing the expression patterns of protein and RNA across developmental stages and tissues. The data are summarized in a single chart for comparison among status and methods.

Thus far, *Ciona* proteomic methods have been successfully used to identify components of sperm cells and to examine their function (see Chapter 14). Another important feature of ascidians as a developmental biological system is that they serve as a model for studying the functions and interactions of gametes (see Chapter 14). The activation of sperm motility and the chemotaxis of sperm to egg during fertilization have been intensively studied in *C. intestinalis* since *Ciona* is an appropriate model for studying chemotaxis and signal transduction during the regulation of cell motility. Furthermore,

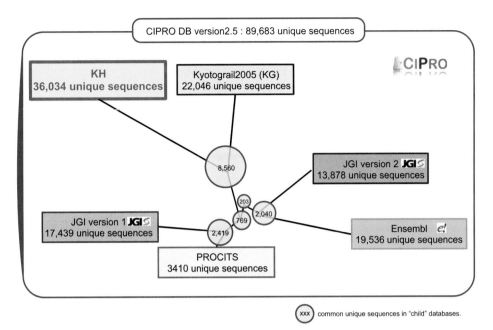

FIGURE 3.5 Similarity and the number of shared sequences among *Ciona intestinalis* proteomes contained in CIPRO database, which contains five proteomes and one local protein data set. Branch lengths represent approximate distances between data set and their internal nodes, based on the proportion of shared sequences with their neighbors. The numbers in cyan circles show the number of sequences shared by descended nodes. (Adapted from Endo *et al.* (2011) *Nucleic Acid Res*, **39**, D807–D814.)

the sperm mitochondria move along the sperm flagella and are excluded from the sperm cell during fertilization. This unique phenomenon, called sperm reaction, contributes to the discrimination of paternal mitochondria during embryonic development.

SELECTED REFERENCES

DEHAL, P., SATOU, Y., CAMPBELL, R.K., CHAPMAN, J. *et al.* (2002) The draft genome of *Ciona intestinalis*: insights into chordate and vertebrate origins. *Science*, **298**, 2157–2167.

ENDO, T., UENO, K., YONEZAWA, K., MINETA, K. *et al.* (2011) CIPRO 2.5: *Ciona intestinalis* protein database, a unique integrated repository of large-scale omics data, bioinformatic analyses and curated annotation, with user rating and reviewing functionality. *Nucleic Acid Res*, **39**, D807–D814.

INABA, K., NOMURA, M., NAKAJIMA, A., HOZUMI, A. *et al.* (2007) Functional proteomics in *Ciona intestinalis*: a breakthrough in the exploration of the molecular and cellular mechanism of ascidian development. *Dev Dyn*, **236**, 1782–1789.

KUSAKABE, T., YOSHIDA, R., KAWAKAMI, I., KUSAKABE, R. *et al.* (2002) Gene exression profiles in tadpole larvae of *Ciona intestinalis*. *Dev Biol*, **242**, 188–203.

PUTNAM, N.H., BUTTS, T., FERRIER, D.E., FURLONG, R.F. *et al.* (2008) The amphioxus genome and the evolution of the chordate karyotype. *Nature*, **453**, 1064–1071.

ROURE, A., ROTHBACHER, U., ROBIN, F., KALMAR, E. *et al.* (2007) A multicassette Gateway vector set for high throughput and comparative analyses in ciona and vertebrate embryos. *PLoS One*, **2**, e916.

SATOH, N. (2003) The ascidian tadpole larva: comparative molecular development and genomics. *Nat Rev Genet*, **4**, 285–295.

SATOU, Y., TAKATORI, N., FUJIWARA, S., NISHIKATA, T. *et al.* (2002) *Ciona intestinalis* cDNA projects: expressed sequence tag analyses and gene expression profiles during embryogenesis. *Gene*, **287**, 83–96.

SATOU, Y., IMAI, K.S., LEVINE, M., KOHARA, Y. *et al.* (2003a) A genomewide survey of developmentally relevant genes in *Ciona intestinalis*: I. Genes for bHLH transcription factors. *Dev Genes Evol*, **213**, 213–2216.

SATOU, Y., KAWASHIMA, T., KOHARA, Y., SATOH, N. (2003b) Large scale EST analyses in *Ciona intestinalis*: its application as Northern blot analyses. *Dev Genes Evol*, **213**, 314–318.

SATOU, Y., HAMAGUCHI, M., TAKEUCHI, K., HASTINGS, K.E. *et al.* (2006) Genomic overview of mRNA 5′-leader *trans*-splicing in the ascidian *Ciona intestinalis*. *Nucleic Acids Res*, **34**, 3378–3388.

SATOU, Y., MINETA, K., OGASAWARA, M., SASAKURA, Y. *et al.* (2008) Improved genome assembly and evidence-based global gene model set for the chordate *Ciona intestinalis*: new insight into intron and operon populations. *Genome Biol*, **9**, R152.

SHOGUCHI, E., HAMAGUCHI, M., SATOH, N. (2008) Genome-wide network of regulatory genes for construction of a chordate embryo. *Dev Biol*, **316**, 498–509.

TASSY, O., DAUGA, D., DAIAN, F., SOBRAL, D. *et al.* (2010) The ANISEED database: digital representation, formalization, and elucidation of a chordate developmental program. *Genome Res*, **20**, 1459–1468.

CHAPTER 4

RESEARCH TOOLS

Compared to other model organisms used in developmental biology, for example, *Drosophila, Caenorhabditis elegans,* and *Xenopus,* it is sometimes difficult to raise marine animal embryos to the juvenile stage, especially in the closed seawater systems of inland laboratories. However, a robust *Ciona* inland culture system has been established to allow for experiments to be performed at any inland laboratory. A number of marine biological laboratories as well as the NBRP of Japan supply *Ciona intestinalis* to researchers year round.

Like other model organisms, many research tools can be applied to ascidians to elucidate the cellular and molecular mechanisms underlying embryogenesis, including the expression and function of developmentally relevant genes and genetic regulatory networks. For example, the comparatively large egg of *Halocynthia roretzi* (∼280 μm in diameter) is advantageous during blastomere isolation and recombination. *Phallusia mammillata* produces optically transparent eggs, which allows for high resolution imaging studies, for example, of ooplasmic segregation. One of the most striking advantages of ascidians is that one can analyze developmental events at the single cell level. WISH reveals the spatial expression of given genes in individual cells. Gene function is easily testable by a variety of means, for example, the microinjection of antisense morpholino oligonucleotides (MOs). Here, the major tools available in ascidian embryology are briefly discussed with a special emphasis on the application of several important techniques to this model system. Recently, a series of protocols for use of *Ciona* embryos appeared, available online from Cold Spring Harbor Laboratory, which are listed under "Protocol References" at the end of this chapter.

4.1 EXPERIMENTAL MANIPULATION OF EMBRYONIC CELLS

As previously mentioned, ascidians were the first animals to have embryonic cells destroyed for the study of embryologic development. Isolation and/or destruction of blastomeres, and combining isolated cells can be easily accomplished after dechorionation. In addition, whole embryos and isolated blastomeres can be treated with pharmacological inhibitors of signaling pathways.

Developmental Genomics of Ascidians, First Edition. Noriyuki Satoh.
© 2014 John Wiley & Sons, Inc. Published 2014 by John Wiley & Sons, Inc.

4.2 TRACING OF CELL DIVISIONS

Developing ascidian embryos are comparatively transparent, and thus allow for the tracing of live cells by imaging techniques. Embryonic cells can also be marked by microinjection of markers such as horseradish peroxidase (HRP) or DiI, and such cells can then easily be traced throughout their lineage. The rapid progress of embryogenesis simplifies the tracing of labeled cells. The configuration of embryonic cells is clearly visible after staining with phalloidin (Fig. 2.1). Three-dimensional (3D) imaging using a confocal microscope allows one to distinguish individual embryonic cells even at later tailbud embryo stages (e.g., Fig. 8.1b).

Recent advances in labeling technology have facilitated more precise cell tracing. The *cis*-regulatory sequence of a gene with the targeted expression pattern is fused to the sequence encoding green fluorescent protein (GFP), red fluorescent protein (RFP), or kaede, and the construct is used to trace cells expressing the target gene of interest. For example, the *Ciona* gene for β-tubulin is specifically expressed in larval CNS cells. A construct of Ci-β-tubulin (promoter)::kaede yields robust kaede expression in the CNS. Kaede-positive cells are isolated and analyzed by microarray to identify genes that are specifically or predominantly expressed in the CNS (see Chapter 10).

The use of photoconvertible fluorescent proteins facilitates tracing of cells at different developmental stages. For example, β-tubulin::kaede is UV-irradiated to convert the marker from green to red. Following CNS cells with photoconverted fluorescent protein elegantly reveals that ependymal cells of the larval CNS function as stem-like cells to produce neurons in the adult nervous system (see Chapter 10). This technique will engender numerous applications in a variety of future developmental biology studies. A fluorescent ubiquitination-based cell cycle indicator (Fucci) has also been applied to *Ciona* embryos to examine the relationship of cell cycle length with the initiation of neurulation morphogenesis.

4.3 ANALYSES OF GENE FUNCTION

4.3.1 Over- and/or Ectopic Expression of Genes

Microinjection of synthetic mRNA into eggs or embryonic cells is a conventional and easy way to test gene function, given that genes *a* and *b* are specifically expressed in embryonic cells A and B, respectively. The function of genes *a* and *b* may be more specifically examined by ectopic expression in cells where they are not usually found. Electroporation (or microinjection) of a construct carrying *a*(promoter)::*b*(ORF) or *b*(promoter)::*a*(ORF) into B or A cells, respectively, leads to ectopic expression of the *b* gene in cell A or the *a* gene in cell B, respectively. The experimental results allow for the more precise interpretation of both genes' function(s).

4.3.2 Knockdown of Gene Expression with Specific Morpholino Oligonucleotides and Other Methods

There have been extensive regulatory gene duplication events in the vertebrate lineage since the time of the last shared ascidian and vertebrate ancestor. Owing to functional redundancies in duplicated genes, it is often tedious to determine the function of individual vertebrate genes. As discussed in Chapter 3, the *Ciona* genome contains a basic set of chordate regulatory genes. Specifically, the TF genes and the SPM genes are present in a state prior to the genome-wide duplication event. The reduced redundancy of the *Ciona* genes simplifies the interpretation of TF and SPM gene function. The combination of a streamlined, nonduplicated genome, and simple cell lineages in *Ciona* provides a powerful tool for the elucidation of gene function, and for discovering gene networks that govern the formation of basic chordate tissues such as the notochord, neural tube, and heart.

Ascidians are among the best systems for morpholino antisense oligonucleotide (MO)-mediated gene interference. Usually MOs target 5′ mRNA translation start site sequences to block translation by preventing ribosomes from binding. MOs also target the exon–intron junctions to block the splicing and alter the splicing patterns. As such, MOs enable functional studies of maternal and zygotic genes in a range of model organisms including ascidians. When two differentially designed MOs against a given gene yield the same phenotype, the interference of gene function is strengthened. It is also desirable to "rescue" the targeted gene function by microinjecting synthetic target mRNA to restore gene expression. Many studies have shown that the MO-induced phenotype recapitulates that of a genetic deletion of the targeted gene (see Chapter 8). Thus, MOs offer a rapid and high throughput approach for chordate functional genomics applications.

A medium-sized screening has been conducted to discover *Ciona* genes with novel developmental functions using an MO strategy. For example, the gene KH.C2.176 encodes a protein with leucine-rich repeats and is similar to the uncharacterized human gene *C10orf11*. A knockdown of its function with specific MOs resulted in developmental effects that were similar to embryos that had lost β-catenin function. The defects in KH.C2.176-knockdown embryos were rescued by the overexpression of a constitutively active form, but not the wild-type version, of β-catenin, suggesting that KH.C2.176 plays a role upstream of β-catenin.

Recently, RNA-based interference (RNAi) has been used to inhibit gene function in worms, flies, and mice. In colonial ascidians, gene function can be inhibited via injecting siRNAs into the blood vasculature. The use of shRNA in *Ciona* embryos has been attempted, and RNAi technology is expected to be applied to the *Ciona* system in the near future. In addition, gene silencing by the Zn-finger nuclease (ZFN) system or transcription activator-like effector nucleases (TALENs) is applicable to the *Ciona* model, which should also facilitate analyses of *Ciona* gene functions in future.

4.4 ANALYSES OF REGULATORY MECHANISMS OF SPECIFIC GENE EXPRESSION

To understand the molecular mechanisms that control specific spatio-temporal expression of regulatory genes, and thereby govern the gene regulatory networks involved in the construction of the chordate body plan, the identification and characterization of *cis*-regulatory modules are essential. *Ciona* provides a highly suitable experimental system for such studies.

4.4.1 Electroporation of Reporter Gene Constructs

In most animal models, reporter constructs fused with lacZ or GFP are microinjected into fertilized eggs, and, depending on the species, results are obtained within a few

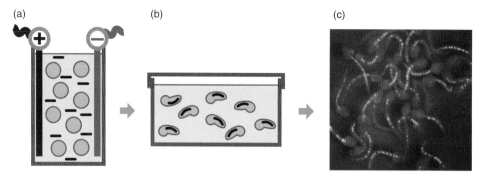

FIGURE 4.1 Schematic showing the electroporation of reporter gene constructs to examine *cis*-regulatory modules governing specific expression of *Ciona* genes. (a) Conventional electroporation of reporter constructs (purple lines) into eggs; (b) cultivation of the embryos to the desired stage; and (**c**) detection of reporter constructs. In this case, a notochord maker is expressed in the cells. (Adapted from cover picture of *Science*, No. 5601, 2002, courtesy of Michael Levine.)

days after injection. With respect to *Ciona*, in addition to ordinal microinjection, reporter constructs can be introduced into fertilized eggs using a simple electroporation method. This permits the simultaneous manipulation of hundreds of synchronously developing embryos (Fig. 4.1). Owing to the fast pace of ascidian embryogenesis, results are obtained within one or two days. In addition, because of the compact organization of genes in the *Ciona* genome (on average there is one gene every 7.7 kb of euchromatic DNA), the core promoter and associated enhancers are usually located within the first 1.5 kb upstream of the transcription start site. In the extreme case of the muscle actin gene, 103 bp of the 5′ flanking sequence are sufficient to drive the muscle-specific expression of reporter constructs.

Ideally, to obtain the most fundamental dataset, the electroporation technique for rapid identification of *cis*-regulatory elements of *Ciona* genes should be applied genome-wide to explore all the regulatory modules. In fact, a medium-sized analysis has already been performed. For example, two genomic domains (>100 kb) containing *Hox2-4* genes and *Hox11/12/13* genes each were discovered. Randomly generated DNA fragments from the 100 kb-long BACs containing the *Hox* genes were inserted into a vector upstream of a minimal promoter and a *lacZ* reporter gene. A total of 222 resultant fusion genes were separately electroporated into fertilized eggs, and their regulatory activities were monitored in larvae. Totally, 21 separable *cis*-regulatory elements were found. In the future, a larger scale, systematic identification of the modules should be carried out on the entire *Ciona* genome.

4.4.2 Phylogenetic Footprinting

Another key strength of the *Ciona* system that facilitates the analysis of gene regulatory mechanisms is that whole-genome sequence assemblies are available for two phylogenetically related *Ciona* species, *C. intestinalis* and *Ciona savignyi*. A comparison of the two allows for the identification of potential regulatory DNAs as conserved noncoding sequences (Fig. 4.2). In addition, sequence polymorphisms within individuals (a remarkable 4.6% in *C. savignyi* and 1.2% in *C. intestinalis*) also assist such studies. These methods have been used to identify a variety of tissue-specific enhancers (see Fig. 6.4).

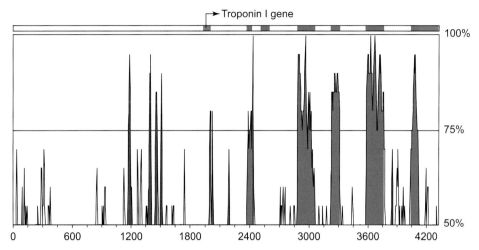

FIGURE 4.2 A Vista alignment between the *C. intestinalis* and *C. savignyi* Troponin I genes and ~2000 bp of 5′ DNA. The *y*-axis shows the percent conservation over a sliding 20 bp window. The *x*-axis shows the position, in bp, of the *C. intestinalis* sequence. Conserved coding sequences are in blue, conserved noncoding sequences are in red. The predicted 5′ and 3′ UTR regions are indicated in green. (Adapted from Satoh *et al.* (2003) *Trends Genet*, **19**, 376–381.)

A web-based database has already been established for the identification of cis-regulatory elements in the 5′ flanking regions of all known tissue-specific genes in the *Ciona* genome (http://dbtgr.hgc.jp/).

4.5   MICROARRAY

With the aid of large, high quality *C. intestinalis* genomic and cDNA datasets, several types of microarrays have been prepared to examine gene expression profiles under normal and experimental conditions. They include Affymetrix GPL15657; Agilent 44k GPL5576; Agilent 44kx4 GPL14686; Agilent 22k GPL4556; and Agilent 1M genome tiling GPL8993 (GEO, http://www.ncbi.nlm.nih.gov/geo/).

As discussed in Section 4.2, in *Ciona*, the combination of microarray approaches with cell labeling techniques allows one to explore gene expression in targeted cell types. For example, a transgenic line has been constructed in which larval brain cells (of the sensory vesicle) are marked by the *Ci-Nut* (promoter):GFP construct. GFP-positive cells are then isolated for examination by microarray. Approximately 500 genes have been shown to be preferentially expressed in the larval brain. Experiments testing the expression and function of these genes are now underway. Another application of this technique has yielded information regarding the GRN of juvenile heart formation. It is discussed in Chapter 13.

4.6   GENETICS, MUTAGENESIS, AND TRANSGENESIS

C. intestinalis and *C. savignyi* are promising organisms for studies of "forward genetics" because they have (i) a short life cycle, lasting 2–3 months from fertilization to

F1 embryos; (ii) a hermaphroditic reproductive system with inducible self-fertilization, allowing for the production of homozygotic F2 mutants within 4–6 months; (iii) a compact genome that offers opportunities to induce mutations leading to visible phenotypes; and (iv) rapid embryogenesis and a morphologically simple tadpole larva allowing for facile detection of mutant phenotypes. Three types of forward genetics, natural, ethylnitrosourea (ENU)-based, and transposon-based insertional mutagenesis, are applicable to the *Ciona* system.

4.6.1 Natural and ENU-Based Mutagenesis

Wild populations of both *C. intestinalis* and *C. savignyi* provide an extensive source of mutants, presumably because of the high allelic polymorphism of individuals within a given population. A variety of mutants including *aimless* have been identified with abnormal phenotypes that affect differentiation, morphogenesis, and larval behavior. The genetic map facilitates the identification of genes responsible for the mutant phenotypes.

ENU-based mutagenic screens have been successfully introduced to identify a number of *C. savignyi* patterning mutants such as *chonmague* and *vagabond*. *Aimless* and *chonmague* are responsible for notochord formation, while *vagabond* is essential for nervous system formation. The functions of these genes are discussed in later sections of this book (see Chapter 9).

4.6.2 Transposon-Based Stable Germ-Line Transgenesis

Transgenesis is a powerful technique to investigate regulatory mechanisms of gene expression and function. Transgenesis is routinely used in many model organisms, including *Drosophila*, *C. elegans*, zebrafish, and mice. As discussed in Section 4.2, transient transgenesis is a prominent technique applied in *Ciona* to investigate *cis*-regulatory modules required for the specific expression of a given regulatory gene. In addition, stable transgenesis—a splendid technique that allows for the creation of useful marker lines—as well as insertional mutagenesis, enhancer trapping, and related methods have been introduced in the *Ciona* system. Two major methods, transposon-mediated transgenesis and I-SceI-mediated transgenesis, have been used to create *Ciona* stable germ-line transgenesis. Here, transposon-mediated transgenesis is discussed in detail.

4.6.2.1 *Transposable Element* Minos   After numerous trials, Sasakura and his colleagues found that the Tc1/*mariner* transposable element *Minos* is active in *C. intestinalis* and *C. savignyi* (Fig. 4.3a). *Minos* is one of the Tc1/*mariner* transposons isolated from *Drosophila hydei*. Similar to most DNA transposons, *Minos* mobilizes, or transposes, by using a "cut and paste" mechanism. *Minos* is small (about 2000 bp) and simple, with two inverted repeats at each end that flank one open reading frame encoding the transposase. Transposase enzymes excise transposons from DNA at the inverted repeats, and integrate the transposons again at specific DNA sites called target sequences. The target sequences are TA dinucleotides. The double-strand breaks generated by transposon excision are rejoined by the endogenous repair system. As a result of the repair, characteristic sequences called footprints remain (typically, 5′-TACTCGTA-3′ or 5′-TACGAGTA-3′). When the *Minos* transposon is microinjected or electroporated into *Ciona* eggs together with transposase mRNA, as many as 37% of the transposon insertions are transmitted to the subsequent generation. The AT-rich *Ciona* genome (63% AT) appears to improve the efficiency of *Minos* activity.

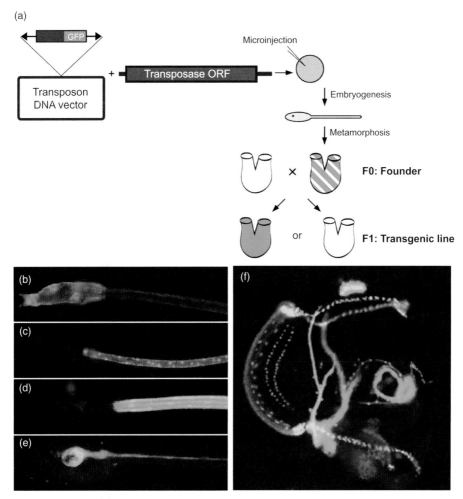

FIGURE 4.3   Transposon-mediated stable transgenesis in *Ciona*. (a) Transgenic line production. When *Minos* (transposon) DNA containing a *GFP* cassette is microinjected into *Ciona* eggs together with transposase mRNA, *Minos/GFP* is excised from the vector DNA and integrated into the *Ciona* genome. Since this event occurs in both somatic cells and germ cells, the transposon-introduced F0 animals are mosaic with respect to their genomes, with only some cells expressing GFP. When *Minos/GFP* is inserted into the chromosomes of primordial germ cells in F0 animals, the animals become founders, and transmit the *Minos/GFP* insertion to the next generation (F1). The F1 animals with *Minos/GFP* insertions are nonmosaic with respect to the insertion, demonstrating a consistent expression of GFP or reporter genes among cells. Once established, *Minos* insertions in the ascidian genome are transmitted stably between generations. (b–f) Transgenic lines provide reproducible tissue-specific marker lines: (b) larval epidermis-specific marker, (c) notochord-specific marker, (d) muscle-specific marker, (e) CNS specific marker, and (f) a juvenile of the transgenic line expressing GFP in the muscle (green) and RFP in the endostyle, peripharyngeal band and retropharyngeal band (red). GFP was driven by the *cis*-element of *Tropoinin I* gene and RFP was driven by the enhancer of *Ci-musashi*. (Courtesy of Sasakura (2007).)

When *Minos* is inserted into the chromosomes of primordial germ cells of F0 animals, called founders, they transmit *Minos* insertions to the next generation (Fig. 4.3a). The frequency of F1 progeny with *Minos* insertions varies among founders (from a few to over 50%). The F1 animals with *Minos* insertions show nonmosaic patterning with respect to the insertion. In other words, the expression of reporter genes is consistent throughout the organism. This nonmosaicism is a major technical advantage of stable transgenesis, allowing researchers to obtain hundreds or even thousands of animals with reporter gene expression that is uniform among cells. Additionally, *Minos* insertions in the ascidian genome are transmitted stably between generations. Several transgenic lines have been maintained over 10 generations.

4.6.2.2 Transposon-Based Mutagenesis

If a *Minos*-transposon inserts into a genomic region that is important for gene function, it generates a specific and stable mutant. One such mutant demonstrates an abnormal metamorphosis phenotype. Metamorphosis is a dramatic event that converts the swimming tadpole larva into a sessile juvenile. Tail absorption takes place very rapidly. One mutant was named the "*swimming juvenile (sj),*" because it undergoes trunk metamorphic events, but continues to swim with an intact tail. This mutant is discussed further in Chapter 8.

4.6.2.3 Marker Lines

Stable marker lines have been established using *Minos*-mediated transgenesis (Fig. 4.3b–f). For example, *Ci-Bra* (*Brachyury* of *C. intestinalis*) is a key regulator for notochord differentiation and is expressed specifically in notochord cells. A 500-bp sequence upstream of *Ci-Bra* drives reporter gene expression that is similar to the endogenous *Ci-Bra* expression. When this region is used to create a *Minos*-based transposable element, a marker line is obtained in which GFP is expressed only in notochord cells of developing embryos (Fig. 4.3c). Similarly, marker lines for epidermis (Fig. 4.3b), muscle (Fig. 4.3d), CNS (Fig. 4.3e), and other larval tissues (Fig. 4.3f) are now available.

4.6.2.4 Enhancer Trapping and Other Forward-Genetic Techniques

Enhancer trapping is an excellent technique that allows for the generation of tissue/organ-specific marker lines to identify endogenous enhancers that are sometimes distally located on genes, and to create insertional mutants based on the expression patterns of candidate causal genes. Owing to the compactness of the *Ciona* genome, *Minos* might have a greater probability of being inserted close to an enhancer. For example, one enhancer trap line reveals the presence of enhancers in the introns of *Ci-Musashi* (Fig. 4.3f).

To enhance the efficiency of transposon-mediated techniques, additional methods hold great potential to further refine experimental studies in the *Ciona* system. The first method worth noting is *Minos* remobilization. The second includes several trapping techniques for highly frequent mutagenesis. And the third procedure is the upstream activating sequence (UAS)–Gal4 system. All techniques are currently used to investigate the regulation of gene expression and function. Remobilization of *Minos* within the *Ciona* genome has been achieved by microinjection of transposase mRNA into embryos whose respective genomes contain tandem *Minos* arrays. This method has been applied to enhancer trappings. A transgenic "mutator" line harboring a tandem array of *Minos* vectors for enhancer trapping has already been created. The UAS–Gal4 system, which enables tissue-specific overexpression of genes of interest, will be a powerful tool for studying gene functions. UAS–Gal4 is active in *Ciona*. By establishing Gal4 driver lines through enhancer trap screening, overexpression and misexpression in any tissue will be possible through the creation of one UAS construct. A *Minos*-Gateway vectors system is also available to facilitate vector construction.

Together with the simple *Ciona* transformation by electroporation, the systematic creation of transformation vectors enables the routine generation of many transgenic lines, which will support the establishment of useful marker lines and insertional mutants. These lines are available through the NBRP of Japan (http://marinebio.nbrp.jp/). Therefore, sophisticated genetic techniques are now available in *Ciona* that will facilitate studies of regulator gene expression and function.

SELECTED REFERENCES

AWAZU, S., SASAKI, A., MATSUOKA, T., SATOH, N. *et al.* (2004) An enhancer trap in the ascidian *Ciona intestinalis* identifies enhancers of its *Musashi* orthologous gene. *Dev Biol*, **275**, 459–472.

BROWN, C., JOHNSON, D.S., SIDOW, A. (2007) Functional architecture and evolution of transcriptional elements that drive gene coexpression. *Science*, **317**, 1557–1560.

KEYS, D.N., LEE, B., DI GREGORIO, A., HARAFUJI, N. *et al.* (2005) A saturation screen for *cis*-acting regulatory DNA in the *Hox* genes of *Ciona intestinalis*. *Proc Natl Acad Sci USA*, **102**, 679–683.

HAMADA, M., SHIMOZONO, N., OHTA, N., SATOU, Y. *et al.* (2011) Expression of neuropeptide- and hormone-encoding genes in the *Ciona intestinalis* larval brain. *Dev Biol*, **352**, 202–214.

HORIE, T., SHINKI, R., OGURA, Y., KUSAKABE, T.G. *et al.* (2011) Ependymal cells of chordate larvae are stem-like cells that form the adult nervous system. *Nature*, **469**, 525–528.

JOLY, J.-S., KANO, S., MATSUOKA, T., AUGER, H. *et al.* (2007) Culture of *Ciona intestinalis* in closed systems. *Dev Dyn*, **236**, 1832–1840.

KAWAI, N., OCHIAI, H., SAKUMA, T., YAMADA, L. *et al.* (2012) Efficient targeted mutagenesis of the chordate *Ciona intestinalis* genomes with zinc-finger nucleases. *Dev Growth Differ*, **54**, 535–545.

NAKAMURA, M.J., TERAI, J., OKUBO, R., HOTTA, K. *et al.* (2012) Three-dimensional anatomy of the *Ciona intestinalis* tailbud embryo at single-cell resolution. *Dev Biol*, **372**, 274–284.

NISHIYAMA, A., FUJIWARA, S. (2008) RNA interference by expressing short hairpin RNA in the *Ciona intestinalis* embryo. *Dev Growth Differ*, **50**, 521–529.

OGURA, Y., SAKAUE-SAWANO, A., NAKAGAWA, M., SATOH, N. *et al.* (2011) Coordination of mitosis and morphogenesis: role of a prolonged G2 phase during chordate neurulation. *Development*, **138**, 577–587.

ROURE, A., ROTHBACHER, U., ROBIN, F., KALMAR, E. *et al.* (2007) A multicassette Gateway vector set for high throughput and comparative analyses in *Ciona* and vertebrate embryos. *PLoS One*, **2**, e916.

SASAKURA, Y., (2007) Germline transgenesis and insertional mutagenesis in the ascidian *Ciona intestinalis*. *Dev Dyn*, **236**, 1758–1767.

SATOH, N., SATOU, Y., DAVIDSON, B., LEVINE, M., (2003) *Ciona intestinalis*: an emerging model for whole-genome analyses. *Trends Genet*, **19**, 376–381.

YAMADA, Y., KOBAYASHI, K., SATOU, Y., SATOH, N. (2005) Microarray analysis of localization of maternal transcripts in eggs and early embryos of the ascidian, *Ciona intestinalis*. *Dev Biol*, **284**, 436–550.

COLD SPRING HARBOR LABORATORY PROTOCOLS FOR *Ciona* EMBRYOGENESIS STUDIES

CHRISTIAEN, L., WAGNER, E., SHI, W., LEVINE, M. (2009a) The sea squirt *Ciona intestinalis*. *Cold Spring Harb Protoc*, **12**. doi: 10.1101/pdb.emo138.

CHRISTIAEN, L., WAGNER, E., SHI, W., LEVINE, M. (2009b) Isolation of sea squirt (*Ciona*) gametes, fertilization, dechorionation, and development. *Cold Spring Harb Protoc*, **12**. doi: 10.1101/pdb.prot5344.

CHRISTIAEN, L., WAGNER, E., SHI, W., LEVINE, M. (2009c) Electroporation of transgenic DNAs in the sea squirt *Ciona*. *Cold Spring Harb Protoc*, **12**. doi: 10.1101/pdb.prot5344.

CHRISTIAEN, L., WAGNER, E., SHI, W., LEVINE, M. (2009d) X-gal staining of electroporated sea squirt (*Ciona*) embryos. *Cold Spring Harb Protoc*, **12**. doi: 10.1101/pdb.prot5344.

CHRISTIAEN, L., WAGNER, E., SHI, W., LEVINE, M. (2009e) Microinjection of morpholino oligos and RNAs in sea squirt (Ciona) embryos *Cold Spring Harb Protoc*, **12**. doi: 10.1101/pdb.prot5344.

CHRISTIAEN, L., WAGNER, E., SHI, W., LEVINE, M. (2009f) Whole-mount in situ hybridization on sea squirt (*Ciona intestinalis*) embryos. *Cold Spring Harb Protoc*, **12**. doi: 10.1101/pdb.prot5344.

CHRISTIAEN, L., WAGNER, E., SHI, W., LEVINE, M. (2009g) Isolation of individual cells and tissues from electroporated sea squirt (Ciona) embryos by fluorescence-activated cell sorting (FACS). *Cold Spring Harb Protoc* **12**. doi: 10.1101/pdb.prot5344.

CHAPTER 5

THE FUNCTION AND REGULATION OF MATERNAL TRANSCRIPTS

In Chapters 2–4, we provided basic information on ascidian embryogenesis, genes and genomics, and research tools. From here forward, we begin to introduce the reader to the deeper questions of developmental biology, that is, the molecular and cellular mechanisms and gene regulatory cascades underlying embryogenesis, and discuss how these issues can be addressed using the *Ciona intestinalis* model system.

5.1 GENE EXPRESSION PROFILES OBSERVED DURING THE CIONA LIFE CYCLE

Before discussing the expression and function of regulatory genes involved in specific developmental processes, we should understand the overall gene expression profile during the *C. intestinalis* life cycle, including the significant portion of mRNAs that are maternally contributed. To this end, the expression profiles for 10,415 genes (65% of the 15,452 *C. intestinalis* genes) have been comprehensively examined at 20 different stages by microarray, including 14 embryonic stages, as well as the larval, juvenile, and 4 adult stages. Based on these results, the genes were categorized into five major gene clusters: maternal, embryonic, embryonic and adult, adult, and stably expressed (Fig. 5.1).

The maternal gene cluster is the largest (39% of genes examined), and includes transcripts that are present in the egg and during early embryonic stages and that decrease as embryogenesis proceeds (Fig. 5.1). The transcript abundances return to initial levels during adulthood because of the development of a mature ovary. Similar results have been obtained in *Drosophila* and *Xenopus*, where maternal transcripts represent ∼30% of the entire repertoire of genes in each species. The embryonic gene cluster, containing 13% of all genes, comprises those that are expressed at high levels during embryogenesis but at low levels in adulthood. Approximately 20% of all genes belong to the embryonic and adult gene cluster, which encompasses genes that are highly expressed during the mid-embryonic stage and maintain these expression levels during adulthood. The adult gene cluster contains 16% of all genes, and contains those that are highly expressed in adulthood but expressed at low levels in the egg and embryonic stages. Finally, the stably expressed gene cluster, claiming 13% of all genes, lists those with relatively unchanging expression throughout the life cycle.

Developmental Genomics of Ascidians, First Edition. Noriyuki Satoh.
© 2014 John Wiley & Sons, Inc. Published 2014 by John Wiley & Sons, Inc.

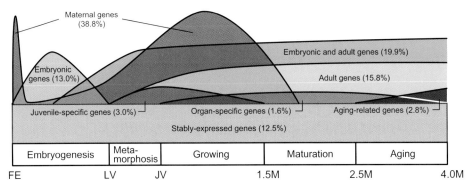

FIGURE 5.1 Pattern diagram depicting the expression profiles of 10,415 genes during the life cycle of *Ciona intestinalis*. There are five major gene clusters, the largest of which is maternally transcribed genes. FE, fertilized eggs; LV, larvae; JV, juveniles. 1.5M, 2.5M, and 4.0M represent 1.5, 2.5, and 4.0 months after metamorphosis, respectively. (Modified from Azumi *et al.* (2007) *Dev Biol*, **308**, 572–582.)

Drosophila genes are expressed in two waves during the life cycle, with embryonic expression patterns recapitulated in pupae, and larval expression patterns recapitulated in adults. By contrast, the expression of most *C. intestinalis* genes changes broadly in one wave over the life cycle. The control mechanisms underlying wide-ranging changes in gene expression should be explored in future studies, along with investigations into the gene expression control at the single gene level, in order to understand embryogenesis as a whole.

5.2 OOCYTE MATURATION, FERTILIZATION, OOPLASMIC SEGREGATION, AND EMBRYONIC AXIS FORMATION

Gametogenesis, oogenesis, and spermatogenesis, are discussed in Chapter 14. As briefly described in Chapter 2, the An–Vg axis is established as the first axis during oogenesis (Fig. 5.2a). The animal pole is defined by the egg position where the polar bodies form, while the vegetal pole is defined as the position opposite the animal pole. Following germinal vesicle breakdown (GVBD), the meiotic spindle migrates to the cortex at the animal pole, and maternal mRNAs, known as postplasmic/PEM RNAs (described later), and mitochondria are excluded from the animal pole region (Fig. 5.2a–c).

Fertilization takes place at the metaphase of meiosis I, which evokes a dynamic reorganization of the zygote cytoplasm and cortical components, traditionally called "ooplasmic segregation." This rearrangement consists of two major phases. The first phase depends on sperm entry that triggers a calcium wave (by a rise in free calcium from 0.5 to 7–10 μM), leading in turn to an actomyosin-driven contraction. The contraction concentrates the rough cortical endoplasmic reticulum (cER)–mRNA domain (discussed later) and the myoplasm, which is cytoplasm rich in mitochondria, in and around a vegetal/contraction pole (Fig. 5.2c, d). The precise localization of the vegetal/contraction pole depends on both the animal–vegetal (An–Vg) axis and the location of sperm entry.

The second phase of reorganization occurs between the completion of meiosis and the first cleavage, and translocates the vegetally localized myoplasm/cER/postplasmic/PEM RNAs to a certain region called the PVC, which is the future posterior pole of the embryo

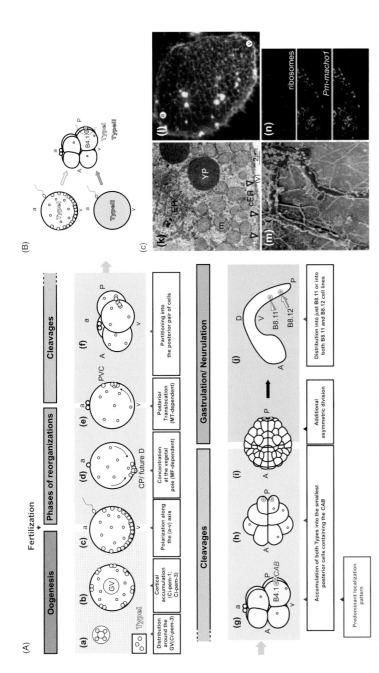

FIGURE 5.2 Schematic representation of postplasmic/PEM RNA distribution from the oocyte to the tailbud stage. Periods of development are indicated in boxes above the respective stages: (A) Distribution patterns for Type I postplasmic/PEM RNAs (yellow circles). (a) Previtellogenic stage I oocytes. (b) Postvitellogenic stage III oocytes. (c) Stage IV oocytes or unfertilized eggs. (d, e) major phases of cytoplasmic reorganization. (f) The 4-cell embryo. (g) The 8-cell embryo. (h) The 16-cell embryo. (i) The 112-cell initial gastrula stage embryo. (j) The tailbud stage embryo. B8.11 and B8.12 cells are indicated by arrowheads and arrows, respectively. (B) After the 8-cell stage, Types I and II postplasmic/PEM RNAs accumulate in the same posterior region, near the CAB. a, animal; v, vegetal; CP/future D, contraction pole/future dorsal pole; A, anterior; P, posterior; D, dorsal; V, ventral. (Modified from Prodon et al. (2007) Dev Dyn, **236**, 1698–1717.) (C) Ultrastructure of cortical ER (cER). (k) Electron microscope section of vegetal region showing cER (red), mitochondria (green), and yolk (YP; blue). (l) Cortical isolation of oocyte showing gradient of the cER network along the A–V axis. (m) Electron microscope deep etch of cER network (red). (n) Fragment of cER network labeled with DiIC16(3) (ribosomes; red), with antisense for Pm-macho1 (RNA; green) showing colocalization of macho1 RNA and cER. (Adapted from Sardet et al. (2007) Dev Dyn, **236**, 1716–1731.)

43

(Fig. 5.2d, e). The dorsal–ventral (D–V) axis of the ascidian embryo corresponds to the An–Vg axis, for which localization of cytoplasmic and cortical domains is established during oogenesis. Thus, the anterior–posterior (A–P) axis is established perpendicular to the An–Vg axis after fertilization. The PVC domains, situated in the periphery of the oocyte, contain developmental determinants and a population of maternal postplasmic/PEM RNAs. Both cER–mRNA and myoplasm domains localized in the posterior region are partitioned equally between the first two blastomeres and then asymmetrically over the next two cleavages (Fig. 5.2f) At the 8-cell stage the cER–mRNA domain compacts and gives rise to a macroscopic cortical structure called the centrosome-attracting body (CAB) (Fig. 5.2g).

The CAB is a macroscopic disc-shaped structure nestled between the plasma membrane and the myoplasm (Fig. 5.3). It attains dimensions of $20 \times 10 \times 5 \ \mu m^3$ in *Ciona*. In following stages, the CAB is responsible for a series of unequal cell divisions of posterior-vegetal blastomeres by attracting a centrosome. This series of unequal cell divisions eventually produce smaller primordial germ cells at the posterior pole (B7.6 cell) (Fig. 2.2). The mechanism of centrosome attraction remains to be fully elucidated, but it is thought that some component of the CAB facilitates the capture of microtubule plus ends, which exert an attractive force on the centrosome. PEM protein encoded by a postplasmic/PEM RNA is required for the attachment of the microtubule bundle to the CAB (Fig. 5.3). The presence of a kinesin-like protein and the polarity proteins, aPKC, PAR-6, and PAR-3 as CAB components provide clues as to possible molecular mechanisms by which this cortical domain may influence microtubule dynamics or anchoring (Fig. 5.3). The postplasmic/PEM RNAs that form part of the CAB are involved in various phenomena of patterning the posterior region of the embryo. A large collection of videos illustrating these reorganization events can be viewed at "Film Archive: Ascidian Eggs and Embryos" at http://biodev.obs-vlfr.fr/recherche/biomarcell/.

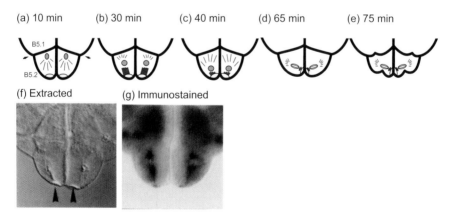

FIGURE 5.3   (a–e) Cartoon illustrations depicting sequential events that occur during the unequal cleavage of the posterior-most blastomere B5.2. Indicated above each drawing is the time after the onset of the previous (fourth) cleavage. The CABs are shown in yellow, the nuclei in blue, and microtubules in red. An assembly of microtubules emanating from the posterior centrosomes focuses on the CABs (b), and then shortens to bring the centrosomes and nuclei close to the CABs (c), causing posteriorly biased positioning of the mitotic apparatus (d, e). (f) Nomarski image of an extracted embryo at the stage corresponding to the drawing in (b). Arrowheads indicate the CABs. (g) α-Tubulin staining of the same stage embryo as in (f). (Adapted from Kumano and Nishida (2007) *Dev Dyn*, **236**, 1732–1747.)

5.3   CONTRIBUTION OF MATERNAL INFORMATION

Since Chabry's experiments in 1887 and Conklin's description in 1905, the ascidian has been regarded as an organism that exhibits a typical mosaic mode of embryogenesis. An enormous contribution of maternally supplied factors (maternal mRNAs/proteins, macromolecular complexes, and cellular components) is required to accomplish ascidian embryogenesis. A series of micromanipulation experiments performed by H. Nishida in the 1990s using *Halocynthia roretzi* embryos revealed that the periphery of the zygote contains three fate determinants for epidermis, endoderm, and muscle, and also contains two morphogenetic process determinants affecting unequal cleavages and gastrulation. For example, Nishida (1994) removed the PVC from a 1-cell embryo and transplanted the PVC into the anterior side of another 1-cell embryo. The donor embryo cleaved equally after the 8-cell stage to form a mirror image blastomere configuration of the anterior half at the vegetal hemisphere. As a result, notochord precursor blastomeres were doubled in both the anterior and posterior-vegetal hemispheres. In other words, the posterior cytoplasm of the ascidian egg contains factor(s) responsible for the unequal cleavage of B4.1 descendents and for the establishment along the A–P axis.

5.4   POSTPLASMIC/PEM RNAs

As previously mentioned, the largest of the five major groups of gene expression clusters in *Ciona* is the maternally transcribed gene cluster. This collection of molecules plays pivotal roles in embryonic axis formation and the specification of embryonic cells. The presence of various maternal mRNAs suggests several different modes of localization within the egg cytoplasm, but this is unlikely. A number of studies have shown that the presence of postplasmic/PEM RNAs at the posterior pole appears to be a predominant localization pattern of maternal transcripts in ascidian eggs and early embryos (Fig. 5.4). To date, ~50 such localized RNAs are known (Table 5.1); most of them were identified from WISH studies that were performed using two ascidian species, *H. roretzi* (MAGEST: http://magest.hgc.jp/) and *C. intestinalis* (cDNA project: http://ghost.zool.kyoto-u.ac.jp/indexr1.html). Two other notable distribution patterns are the mitochondria-like and the uniform cortical localization patterns.

5.4.1   Two Types of Postplasmic/PEM RNAs

Postplasmic/PEM mRNAs are subdivided into two types, according to their mode of localization in unfertilized eggs, redistribution between fertilization and the 8-cell stage, and abundance (Fig. 5.2B). As originally found with *pem* RNA (Fig. 5.4), the abundant Type I postplasmic/PEM RNAs demonstrate a polarized cortical distribution along the An–Vg axis in unfertilized eggs (Fig. 5.2A and Fig. 5.4a). They are transiently concentrated in the vegetal/contraction pole after the first phase of cytoplasmic reorganization (Fig. 5.2c, d and Fig. 5.4b), and are subsequently translocated to the PVC after the second phase of cytoplasmic reorganization (Fig. 5.2e and Fig. 5.4c). After the first cleavage, Type I postplasmic/PEM RNAs partition equally into blastomeres (Fig. 5.2f and Fig. 5.4d). Then, after the second cleavage, Type I postplasmic/PEM RNAs segregate into the posterior pair of cells (Fig. 5.2f and Fig. 5.4e), and finally, at the 8-cell stage, concentrate in the posterior-most region of B4.1 (Fig. 5.2g and Fig. 5.4f). Nonetheless, some Type I postplasmic/PEM RNAs are also present in blastomeres other than B4.1.

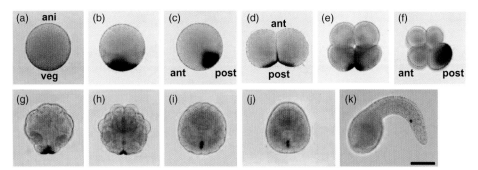

FIGURE 5.4 The distribution of *pem* maternal mRNA, as detected by WISH, marks the posterior end of developing embryos. (a) An unfertilized egg, lateral view. ani, animal pole of the egg; veg, vegetal pole of the egg. (b, c) A fertilized egg after completion of the first (b) and second phase (c) of ooplasmic segregation, lateral view. ant, anterior side of the egg; post, posterior side of the egg. (d) 2-cell embryo, animal pole view. (e) 4-cell embryo, animal pole view. (f) 8-cell embryo, lateral view. (g) 16-cell embryo, vegetal pole view. (h) 32-cell embryo, vegetal pole view. (i) Early gastrula, vegetal pole view. (j) Embryo at the neural plate stage, vegetal pole view. (k) A tailbud embryo, side view. Scale bar represents 100 μm. (Adapted from Yoshida *et al.* (1996) *Development*, **122**, 2005–2012.)

Some Type I postplasmic/PEM RNAs including *pem* and *macho-1* are bound to a network of rough cER tethered to the plasma membrane of the oocyte, thus forming a cER–mRNA domain (Fig. 5.2C). This cER–mRNA domain is relocated after fertilization and concentrates into the CAB at the 8-cell stage. Type I postplasmic/PEM RNAs are observed in a number of ascidian species, including *C. intestinalis, C. savignyi, H. roretzi*, and *Phallusia mammillata*. They encode proteins with diverse functions as discussed later (Table 5.1).

On the other hand, Type II postplasmic/PEM RNAs are first distributed uniformly throughout the egg cytoplasm and then begin to accumulate in the posterior region at about the 4-cell stage (Fig. 5.2B). The mechanism of relocalization remains to be elucidated, although experiments using cytoskeletal inhibitors suggest that distinct mechanisms are involved in the posterior localization of Types I and II postplasmic/PEM RNAs to the CAB region. However, there are several exceptions in the two-types categorization of postplasmic/PEM RNAs. Further studies may define this problem by elucidating cellular and molecular mechanisms underlying each of the postplasmic/PEM RNAs.

5.4.2 Possible Functions of Postplasmic/PEM RNAs

Examples of postplasmic/PEM RNAs include *pem, macho-1, POPK-1, wnt-5*, and *pem-3* (Table 5.1). *pem* encodes a protein with a C-terminal Groucho-like repression motif (WPRW; Try-Arg-Pro-Try). The function of *pem-1* in relation to silencing of zygotically expressed genes in germ-line cells is discussed in Chapter 14. *Macho-1* encodes a Zic family member zinc-finger transcription factor that acts as a muscle determinant. The role of *macho-1* with respect to muscle cell differentiation is discussed in Chapter 6. The functions of several other postplasmic/PEM RNAs are briefly addressed in the following paragraphs.

5.4.2.1 POPK-1 *Hr-POPK-1* encodes a Ser/Thr kinase protein with strong similarities to the *Sad-1* gene in *Caenorhabditis elegans*. *Hr-POPK-1* acts upstream of *macho-1*.

TABLE 5.1   Representatives of Postplasmic/PEM Genes of *Ciona intestinalis*, *Ciona savignyi*, and *Halocynthia roretzi*

Gene	Gene name	Type[a]	Expression tissues expressed zygotically[b]		Function/ characterization
pem	*Ci-pem-1*	Type I	M		It plays roles in anterodorsal patterning, unequal cell division, and germ cell formation
	Cs-pem-1	Type I	M		
	Hr-pem-1	Type I	M		
macho1	*Ci-macho-1*	Type I	MZ	NS	A zinc-finger transcriptional factor. It acts as a muscle determinant and a posterior patterning determinant responsible for mesenchyme formation
	Cs-macho-1	Type I	MZ	NS	
	Hr-macho-1	Type I	MZ	NS	
POPK1	*Ci-POPK-1*	Type I	MZ	Ep	A Ser/Thr kinase protein similar to *Caenorhabditis elegans* Sad-1 gene. It is upstream of macho-1 and plays a role in concentration of the cortical ER, subsequent localization of Type I postplasmic/PEM mRNAs, and concentration of putative germinal granules
	Cs-POPK-1	n.e.	MZ	Ep	
	Hr-POPK-1	Type I	MZ	EP, NS	
Wnt5	*Ci-wnt-5*	Type I	MZ	Ep, Mu	Secreted intercellular signaling protein. It plays roles in the primary notochord, mesencyme, and muscle cell formation. Upstream of macho-1
	Cs-wnt-5	n.e.	MZ	Ep, Mu	
	Hr-wnt-5	Type I	MZ	Ep, Mu, No, En, Me, NS	
pem-3	*Ci-pem-3*	Type I	MZ	NS, Me, ES	A RNA-binding protein with KH and RING finger domain, similar to *C. elegans* MEX-3. It plays a role in the differentiation of the brain of the larvae
	Cs-pem-3	Type I	MZ	NS, Me, ES	
	Hr-pem-3	Type I	MZ	NS, Me	

(continued)

TABLE 5.1 (*Continued*)

Gene	Gene name	Type[a]	Expression tissues expressed zygotically[b]		Function/ characterization
pem-2	*Ci-pem-2*	Type I	MZ	NS, Mu	Containing a SH3 and a GEF domain
	Cs-pem-2	n.e.	M		
ZF1	*Ci-ZF-1*	Type I	M		A C3H-type zinc-finger protein similar to *C. elegans* PIE-1
	Cs-ZF-1	n.e.	M		
	Hr-ZF-1	Type I	M		
VH	*Ci-VH*	Type II	MZ	En	A Vasa ortholog
	Cs-VH	n.e.	MZ	En	
Dll-B	*Ci-Dll-B*	Type I	MZ	NS, Ep	A homeobox transcriptional factor
	Cs-Dll-B	n.e.	MZ	NS, Ep	
Prd-B	*Ci-prd-B*	n.d.	MZ	Me, Mu	A homeobox transcriptional factor
	Cs-prd-B		n.e.		
prep	*Ci-prep*	n.d.	MZ	Me	A homeobox transcriptional factor
	Cs-prep	n.e.	n.e.		
LAG1-like3	*Ci-LAG1-like3*	n.d.	MZ	Ep	A homeobox transcriptional factor
ets/pointed2	*Ci-ets/pointed2*		MZ	NS, Ep	A ETS-domain transcription factor, target of FGF/MAPK signaling. It is required for induction of notochord, mesenchyme, and brain
	Cs-ets/pointed2		n.e.		
	Hr-ets	Type II	MZ	NS, Ep, No	
Fli/ERG4	*Ci-FLI/ERG4*	n.d.	MZ	En, No	A ETS-domain transcription factor
FoxJ2	*Ci-foxJ2*	n.d.	M		A Fox transcription factor
	Cs-foxJ2	n.e.	n.e.		
scalloped/TEF1	*Ci-scalloped/tef1*	n.d.	MZ	NS	A transcription factor similar to scalloped and TEF
	Cs-scalloped/tef1		n.e.		
Eph1	*Ci-eph-1*	Type I	M		A tyrosine kinase receptor for Ephyrins, beta-catenin downstream gene
	Cs-eph-1	n.e.	M		
Eph2	*Ci-eph-2*	Type I	MZ	NS, Ep	A tyrosine kinase receptor for Ephyrins

TABLE 5.1   (*Continued*)

Gene	Gene name	Type[a]	Expression tissues expressed zygotically[b]		Function/ characterization
GCNF	*Ci-GCNF*	Type I	MZ	NS, Me, Ep	A nuclear receptor
	Cs-GCNF	n.e.	MZ	NS, Me, Ep	
Tolloid	*Ci-tolloid*	Type I	MZ	NS, Me, En	A TGF-β signal transduction molecule

n.d., not determined.

n.e., not examined.

M, maternally expressed.

MZ, maternally and zygotically expressed.

[a] Classification of genes depending on criteria described in Sasakura *et al.* (2000).

[b] NS, nervous system; Me, mesencyme; Ep, epidermis; Mu, Muscle, En, endoderm; ES, endodermal strand; No, notochord.

Suppression of *Hr-POPK-1* function by MO leads to significant defects in the translocation of Type I postplasmic/PEM RNAs, CAB formation, posterior unequal cleavages, and the formation of muscle and mesenchymal tissues. As in *C. elegans, Sad-1* plays a role in neuronal presynaptic vesicle clustering, the ascidian POPK-1/Sad-1 kinase is likely to also be involved in the trafficking of membrane components, and thus may affect the translocation of Type I postplasmic/PEM RNAs.

5.4.2.2   wnt-5   Wnt is a secreted intercellular signaling molecule. In *H. roretzi*, suppressing *Hr-wnt-5* function results in failure of notochord-specific *Hr-Bra* expression by abrogation of proper separation of endoderm and mesoderm germ layers. *Hr-wnt-5* is also likely to be involved in the formation of mesenchyme and muscle.

5.4.2.3   pem-3   *Cs-pem-3* encodes an RNA-binding protein. The KH domains (RNA-binding modules) of pem-3 are highly homologous to those of the *C. elegans* protein, MEX-3 (Muscle Excess protein-3), and the mammalian protein, TINO. In *C. elegans*, MEX-3 is involved in establishing the anterior–posterior asymmetry of the embryo and is a component of the germ line-specific P granules. The role of maternal *Cs-pem-3* has not been adequately addressed, but its zygotic expression is likely to be involved in the differentiation of the ascidian larval brain.

5.4.3   Localization Mechanisms of Postplasmic/PEM RNAs

How is the specific localization of postplasmic/PEM RNAs controlled? At least two components are involved. One is the subcellular structure that anchors the postplasmic/PEM RNAs to the posterior-most region of the egg and embryo. The other involves elements that are present within the postplasmic/PEM RNAs themselves.

5.4.3.1   *Cytoplasmic Components for Localization: CAB*   As mentioned above, the CAB is a macromolecular cortical ER structure rich in ER that forms in the posterior-most cytoplasm of B4.1 (Fig. 5.3). Some postplasmic/PEM RNAs are suggested to be anchored to the cortical ER. The CAB is responsible for three successive asymmetric divisions at an embryo's posterior pole from the 8-cell stage to the 64-cell stage, ensuring the faithful partitioning of postplasmic/PEM RNAs into the daughter cells B5.2, B6.3, and B7.6 after each round of cell division.

5.4.3.2   *Molecular Components for Localization: 3′UTR of RNA*   The 3′UTRs untranslated regions of *Hr-wnt-5, Hr-POPK-1, Hr-ZF-1, Ci-pem, Ci-wnt-5, Ci-ZF-1, Hr-pet-1, Hr-pet-2, and Hr-pet-3* are necessary and sufficient to direct the trafficking of these mRNAs to the posterior pole. The former six belong to the Type I postplasmic/PEM RNA group and the latter three to the Type II postplasmic/PEM RNA group. Cross-species experiments testing localization mechanisms suggest the presence of conserved 3′UTR elements among *C. intestinalis, C. savignyi*, and *H. roretzi*.

The 3′UTR of *Hr-wnt-5* and *Hr-POPK-1* mRNAs contain UG dinucleotide–repetitive elements (UGREs) consisting of small nucleotide sequences (20–25 nucleotides) necessary to drive posterior localization. By comparing the 3′UTR sequences of two *Ciona pem* mRNAs, two highly conserved regions were identified. Both regions contain a CAC-repeat within a uracil-rich region (e.g., 5′-ucuCACcu-3′). The deletion of this sequence drastically reduces the efficiency of mRNA localization. Similar CAC-containing repeats are abundant in the 3′UTRs of many postplasmic/PEM RNAs. The high density clusters of short CAC-containing motifs characterize the localization elements of virtually all mRNAs confined to the vegetal cortex of *Xenopus laevis* oocytes. Therefore, the conserved CAC-motifs, in concert with the uracil-rich UGRE elements, are likely to play crucial roles in the posterior localization. The 3′UTRs of postplasmic/PEM RNAs are also involved in mRNA translational repression.

By contrast to the Type I postplasmic/PEM RNAs, an analysis of the 3′UTR of a Type II postplasmic/PEM RNA, *Hr-pet-1*, suggests that it does not contain translational repression activity. Future studies may address whether the translational repression of Type I postplasmic/PEM RNAs also occurs in immature oocytes, and whether there is a conserved mechanism for translational repression in different ascidian species.

It is a common developmental theme for maternal mRNAs, specifically localized within the egg cytoplasm, to play pivotal roles in embryonic axis formation and cell specification. Cellular and molecular mechanisms involved in the localization of these mRNAs have been studied intensively in *Drosophila* and *C. elegans*. By contrast, in chordates, the mechanisms largely remain to be elucidated, although experiments in *Xenopus* have yielded some information. Completing the picture of how maternal mRNAs are localized and function in chordates will be an important task, requiring ever more sophisticated techniques. Accordingly, ascidians represent an attractive model system for such studies.

SELECTED REFERENCES

AZUMI, K., SABAU, S.V., FUJIE, M., USAMI, T. *et al.* (2007) Gene expression profile during the life cycle of the urochordate *Ciona intestinalis. Dev Biol*, **308**, 572–582.

KUMANO, G., NISHIDA, H. (2007) Ascidian embryonic development: an emerging model system for the study of cell fate specification in chordates. *Dev Dyn*, **236**, 1732–1747.

Makabe, K.W., Kawashima, T., Kawashima, S., Minokawa, T. *et al.* (2001) Large-scale cDNA analysis of the maternal genetic information in the egg of the ascidian, *Halocynthia roretzi*, for the gene expression catalog during development. *Development*, **128**, 2555–2567.

Makabe, K.W., Nishida, H. (2012) Cytoplasmic localization and reorganization in ascidian eggs: role of postplasmic/PEM RNAs in axis formation and fate determination. *WIREs Dev Biol*, **1**, 501–518.

Nakamura, Y., Makabe, K.W., Nishida, H. (2006) The functional analysis of Type I postplasmic/PEM mRNAs in embryos of the ascidian Halocynthia roretzi. *Dev Genes Evol*, **216**, 69–80.

Nakamura, Y., Makabe, K.W., Nishida, H. (2005) POPK-1/Sad-1 kinase is required for the proper transloca-tion of maternal mRNAs and putative germ plasm at the posterior pole of the ascidian embryo. *Development*, **132**, 4731–4742.

Nishida, H. (1994) Localization of determinants for formation of the anterior-posterior axis in eggs of the ascidian *Halocynthia roretzi*. *Development*, **120**, 3093–3104.

Nishida, H. (2005) Specification of embryonic axis and mosaic development in ascidians. *Dev Dyn*, **233**, 1177–1193.

Nishikata, T., Hibino, T., Nishida, H. (1999) The centrosome-attracting body, micro-tubule system, and posterior egg cytoplasm are involved in positioning of cleavage planes in the ascidian embryo. *Dev Biol*, **209**, 72–85.

Prodon, F., Dru, P., Roegiers, F., Sardet, C. (2005). Polarity of the ascidian egg cortex and relocalization of cER and mRNAs in the early embryo. *J Cell Sci*, **118**, 2393–2404.

Prodon, F., Yamada, L., Shirae-Kurabayashi, M. *et al.* (2007) Postplasmic/PEM RNAs: a class of localized maternal mRNAs with multiple roles in cell polarity and development in ascidian embryos. *Dev Dyn*, **236**, 1698–1715.

Sardet, C., Paix, A., Prodon, F., Chenevert, J. (2007) From oocyte to 16-cell stage: cytoplasmic and cortical reorganizations that pattern the ascidian embryo. *Dev Dyn*, **236**, 1716–1731.

Sasakura, Y., Ogasawara, M., Makabe, K.W. (2000). Two pathways of maternal RNA localization at the posterior-vegetal cytoplasm in early ascidian embryo. *Dev Biol*, **220**, 365–378.

Sasakura, Y., Makabe, K.W. (2002). Identification of *cis* elements which direct the localization of maternal mRNAs to the posterior pole of ascidian embryos. *Dev Biol*, **250**, 128–144.

Satou, Y. (1999) *posterior end mark 3 (pem-3)*, an ascidian maternally expressed gene with localized mRNA encodes a protein with *Caenorhabditis elegans* MEX-3-like KH domains. *Dev Biol*, **212**, 337–350.

Yamada, L., Kobayashi, K., Satou, Y., Satoh, N. (2005) Microarray analysis of localization of maternal transcripts in eggs and early embryos of the ascidian, *Ciona intestinalis*. *Dev Biol*, **284**, 536–550.

Yamada, L. (2006) Embryonic expression pro-files and conserved localization mechanisms of pem/postplasmic mRNAs of two species of ascidian, *Ciona intestinalis* and *Ciona savignyi*. *Dev Biol*, **296**, 524–536.

Yoshida, S., Marikawa, Y., Satoh, N. (1996) *posterior end mark*, a novel maternal gene encoding a localized factor in the ascidian embryo. *Development*, **122**, 2005–2012.

CHAPTER 6

LARVAL TAIL MUSCLE

In 1905, Conklin found that a specific, yellowish cytoplasm, called myoplasm, is present in fertilized eggs of *Cynthia* (*Styela*) *partita*, segregated into a posterior cytoplasm during ooplasmic segregation, and partitioned into the primary muscle cell lineage (B-line) during cleavage stage. He speculated that the myoplasm contained a maternal factor essential for the differentiation of larval tail muscle cells. Since then, the ascidian embryo has served as an experimental system to explore the cellular and molecular mechanisms underlying the autonomous specification of embryonic cells. The discovery of a muscle determinant gene, *macho-1*, by Nishida and Sawada (2001) was one of the triumphs of molecular developmental biology. At the same time, the secondary lineage (A- and b-line) muscle cells provide an excellent model to explore the evolutionary alteration of regulatory genes.

6.1 PRIMARY (B-LINE) MUSCLE CELLS: AUTONOMOUS SPECIFICATION AND GENE REGULATORY NETWORKS

The developmental potential that allows B4.1 to give rise to muscle is inherited by B5.1 and B5.2 of the 16-cell embryo, and by B6.2, B6.3, and B6.4 of the 32-cell embryo (Fig. 6.1a, b). At the 64-cell stage, B7.4, B7.5, and B7.8 are the muscle precursors, and the developmental fate of B7.4 and B7.8 is restricted to muscle. B7.4 and B7.8 give rise to two clones of eight and four muscle cells on each side of the tail after three and two divisions, respectively (Fig. 6.1a, b). Meanwhile, B7.5 is destined to give rise to four muscle cells of the anterior larval tail and also to TVCs, the latter being juvenile heart muscle precursors (see Chapter 13).

Aforementioned, the ascidian primary muscle cell is a well-known example of the autonomous specification of embryonic cells in metazoans. Autonomous differentiation has been demonstrated in two ways. First, through the observation that the myoplasm with muscle-forming potential is segregated into B-line muscle cells in both normal and cleavage-arrested embryos. Second, by showing that B4.1 cells isolated from the 8-cell embryo can differentiate into muscle cells in the absence of cues from the other blastomeres. Additional support for the presence of muscle determinant in the myoplasm has come from experiments where alteration of the myoplasm distribution led to the development of extra muscle cells. Furthermore, in studies where a myoplasm-containing fragment of B4.1 was fused to a4.2 (the sole nonmuscle forming cell of the 8-cell embryo) muscle differentiation was observed in the partial embryos that developed from the manipulated a4.2.

Developmental Genomics of Ascidians, First Edition. Noriyuki Satoh.
© 2014 John Wiley & Sons, Inc. Published 2014 by John Wiley & Sons, Inc.

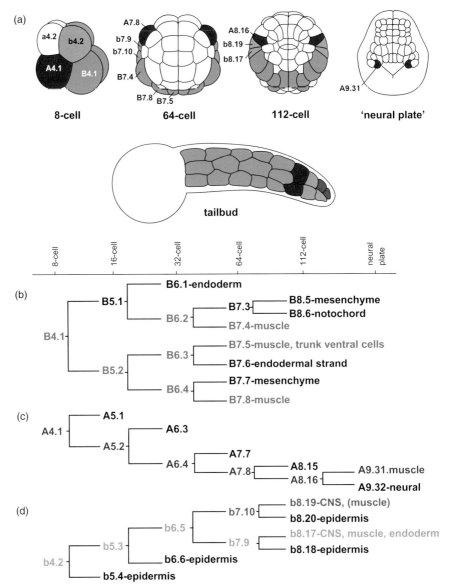

FIGURE 6.1 Lineage of primary (B-line) and secondary (A- and b-line) muscle cells in ascidian embryos. (a) Schematics of the 8-cell, 64-cell, 112-cell, neural-plate stage embryo, and tailbud embryo. B-line, green; A-line, red; b-line, orange and brown. (b–d) Lineage diagram from the 8-cell stage to the neural-plate stage. (b) B-line, (c) A-line, and (d) b-line. An additional b7.10-line muscle precursor in *Halocynthia* embryos is shown in dark brown. (Modified from Hudson and Yasuo (2008) *Biol Cell*, **100**, 265–277.)

6.1.1 *macho-1* as a Muscle Determinant Gene

Ultrastructurally the myoplasm is composed of pigment granules, an aggregation of mitochondria, yolk particles, endoplasmic reticulum, fine granular materials, and a cytoskeletal framework (see Fig. 5.2k). The abundance of mitochondria in the myoplasm is accounted for by the fact that mitochondria serve as energy sources for the muscle cells that drive larval locomotion. Until the 1990s, it was uncertain whether the muscle determinant was maternal mRNA or protein or both. Initial answers came from an egg fragmentation experiment. Centrifugation of *Ciona savigyi* eggs yields four different fragment types: nucleated red fragments, and nonnucleated brown, yellow, and black fragments. Only the black fragments contain muscle determinant. As UV irradiation of the black fragment at a wavelength that would affect nucleic acids but not proteins abolished the muscle cell determinant potential, and the black fragments contain no genomic DNA, the determinant is likely to be maternal mRNA. Subtraction of the poly(A)$^+$ RNA found in the red fragments from that found in the black fragments led to the isolation of a specific mRNA with localization that is restricted to the posterior end of early embryos. This gene was named *pem* (*posterior end mark*), and represented the first identification of localized maternal mRNA in ascidians (Chapter 5). *pem* has no role as a muscle determinant but the gene plays pivotal roles in patterning of the ascidian embryo.

macho-1 of *Halocynthia roretzi* is a gene with maternal transcripts that are localized in a similar fashion to *pem* (Fig. 6.2b–g). *macho-1* encodes a Zic family zinc-finger transcription factor and acts as a muscle determinant (Fig. 6.2a). Functional suppression of *macho-1* results in the failure of primary muscle cell formation (Fig. 6.2h, i), and overexpression of the gene by injection of synthetic *macho-1* mRNA promotes the ectopic formation of muscle cells (Fig. 6.2j, k). Therefore, *macho-1* is the localized factor that is both necessary and sufficient for specification of muscle fate. In time, additional functions of *macho-1* have become clear in *Halocynthia* embryos. Namely, macho-1 plays a central role in modulating responsiveness to the inductive signals of fibroblast growth factor (FGF) in notochord and mesenchyme cell induction, indicating that macho-1 also acts as a posterior patterning factor responsible for mesenchyme and muscle formation.

6.1.2 Genetic Cascade of Muscle Differentiation

Orthologs of *macho-1* have been isolated in *Ciona intestinalis* and *C. savignyi*. *Ciona macho-1* genes are expressed both maternally and zygotically, and the zygotic transcript is produced in cells of the CNS. Vertebrate members of the Zic family are also expressed in and function in the CNS. Since, to date, *macho-1* has been found only in ascidians, it is likely that ascidians recruited this Zic-family member as a muscle determinant during their evolution.

Maternal *Ci-macho1* activates *Tbx6b* and *Tbx6c* at the 16-cell stage (Fig. 6.3). Then, at the 64-cell stage, *Tbx6b*, *Tbx6c*, and *ZicL* likely activate the bHLH myogenic factor *MyoD*, which plays a key role in the transcriptional activation of muscle structural genes (Fig. 6.3), suggesting that a myogenic regulatory factor (MRF)-dependent network existed in the tunicate/vertebrate shared ancestor. *Tbx6b* and *Tbx6c* are likely to exert cross-regulatory interactions as part of their regulatory functions (Fig. 6.3).

6.1.3 Regulation of Muscle Structural Gene Expression

The larval tail muscle is composed of unicellular striated cells, and differentiated muscle cells contain multiple striated myofibrils consisting of thick and thin myofilaments that occur as a single layer in the periphery of the muscle cell. Myofilaments are first detected

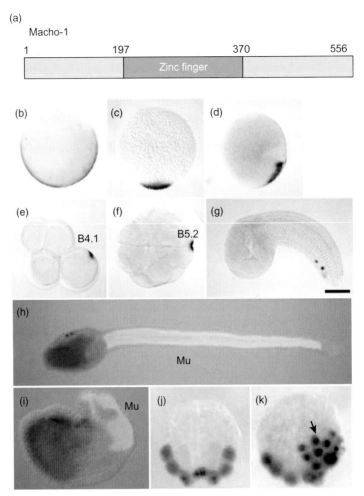

FIGURE 6.2 Macho-1 as a larval tail muscle determinant in *Halocynthia* embryos. (a) Structure of macho-1 protein. The cDNA clone, 2,210 bp in length, encodes a 556-amino acid protein with DNA-binding domain of five zinc fingers in the central part. (b–g) Localization of *macho-1* mRNA. Anterior is to the left. (b) Unfertilized egg. (c, d) Fertilized egg after the first (c) and second phase (d) of ooplasmic movements. (e) The 8-cell stage, lateral view. *In situ* hybridization signal is localized to tiny spots in B4.1 blastomeres. (f) The 16-cell stage and (g) early tailbud stage. Scale bar, 50 μm. (h) Control larva injected with scrambled oligonucleotide and stained with antimyosin antibody (green) and for endoderm-specific alkaline phosphatase (ALP) activity (brown). (i) Experimental larva injected with antisense oligonucleotide showing reduced amount of myosin-positive cells (A- and b-line secondary muscle cells are not affected). (j) In control embryo at the 110-cell stage showing actin expression in 10 precursor blastomeres of primary muscle cells. (k) An embryo injected with synthesized *macho-1* mRNA at the 110-cell stage showing ectopic actin gene expression (arrow). (Adapted from Nishida and Sawada (2001) *Nature*, **409**, 724–728.)

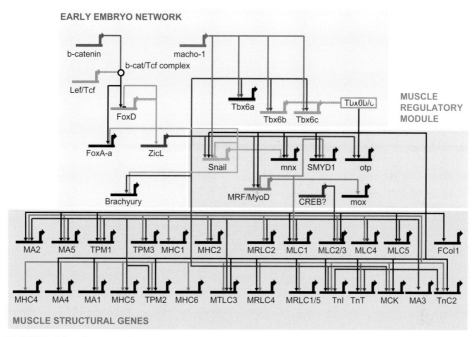

FIGURE 6.3 Gene regulatory network (GRN) involved in specification and differentiation of B-line muscle cells in *Ciona* embryos. Constructed by compiling data from various sources. (Modified from Wang and Christiaen (2012) *Curr Top Dev Biol*, **98**, 147–172.)

at the middle of the tailbud stage in *Ciona* embryos. A characteristic feature of muscle cell differentiation in ascidian embryos is very early expression of muscle structural genes including muscle-type actin, myosin heavy chain, troponin T, troponin I, and tropomyosin. Transcripts of the genes are detectable by WISH as early as the ∼64-cell stage, although the proteins themselves appear during gastrulation. This suggests that if MyoD acts as a regulator, it activates structural gene expression as soon as their protein products are produced.

A comprehensive computational and experimental approach in both *C. intestinalis* and *C. savignyi* toward identifying the upstream motifs responsible for the muscle-specific expression of genes encoding actin, troponin C, troponin T, tropomyosin, myosin essential light chain, myosin regulatory light chain, and myosin heavy chain defined three essential motifs. One motif contains an E-box (CAGVTG), and may be targeted by myogenic factors (Fig. 6.4a, Fig. 6.4b green circles). Another motif contains a consensus-binding site (TGACGTCA) for a cyclic AMP response-element (CRE) binding transcription factor (CREB) (Fig. 6.4a, Fig. 6.4b red circles). This motif is distributed preferentially in regions from −200 to −100 bp relative to the translational start sites (Fig. 6.4b), although it remains to be determined whether CREB plays a role in muscle-specific structural gene expression. The third motif is associated with Tbx binding site (GWTCACACCT) (Fig. 6.4b blue circles), suggesting a direct activation of muscle-specific genes by Tbx6b/Tbx6c. Although further analyses will be required, it is evident that the ascidian primary muscle cell lineage provides a system to completely describe the gene regulatory network responsible for differentiation.

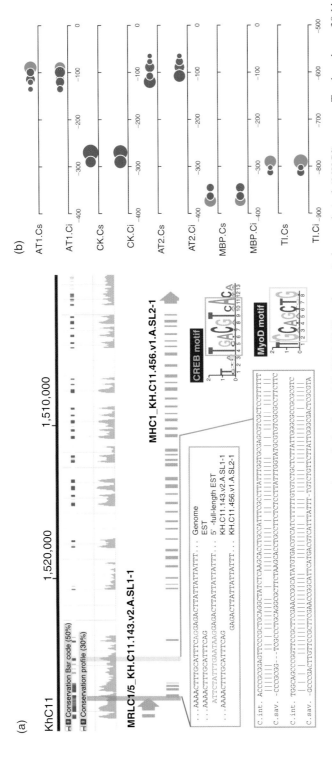

FIGURE 6.4 (a) Genomic organization of *myosin regulatory light chanin 1/5* (*MRLC1/5*) and *myosin heavy chain 1* (*MHC1*) operon. Top bar shows 30-kbp region (from 1500 to 1530 kb) of the *Ciona intestinalis* genome KhC11 scaffold. Green boxes indicate exons and green solid arrows indicate the last exons. The last exon of *MRLC1/5* abuts the first exon of *MHC1* (blue box), showing a typical operon. The junction contains a splice acceptor site (red AG) preceded by a polypyrimidine tract and the 5′ full-length EST, for *MHC1* contains a 5′ SL leader sequence (red) indicative of SL trans-splicing. A conserved noncoding region is located in the fourth intron of *MHC1* (red box). This sequence contains short, conserved blocks almost identical to the overexpressed motifs identified upstream of muscle genes. These motifs correspond to possible binding sites for CREB (orange box) and MRF/MyoD (yellow box). (Adapted from Wang and Christiaen (2012) *Curr Top Dev Biol*, 98, 147–172.) (b) Distribution of *cis-*regulatory motifs at the muscle-related 10 loci of *C. intestinalis* (Ci) and *C. savignyi* (Cs). AT1, α-tropomyosin 1; CK, creatine kinase; AT2, α-tropomyosin 2; MBP, myosin binding protein; and TI, troponin I. Motifs are depicted as circles, and color indicates motif type: CRE (red), MyoD (green), and Tbx6 (blue). Labels below axes indicate distance to transcription start site. Area of circle is proportional to estimated motif activity. (Adapted from Brown *et al.* (2007) *Science*, **317**, 1557–1560.)

6.2   SECONDARY (A- AND b-LINE) MUSCLE CELLS: EVOLUTIONARY ALTERATION OF GENE REGULATORY NETWORKS

The evolutionary modification of the metazoan body plan is thought to have been achieved through changes in embryonic cell behavior, including division, positioning, fate restriction, and differentiation, driven in part by alterations in the GRN. Two alternative developmental mechanisms of evolutionary modification are (i) gradual and step-by-step changes, or (ii) drastic, threshold-like alterations in developmental events. A comparison between the molecular mechanisms involved in the secondary lineage muscle cell differentiation of *Ciona* and *Halocynthia* highlights an interesting example with respect to evolutionary changes in the GRN.

6.2.1   A-Line: Specification Mechanisms

In both *Ciona* and *Halocynthia*, development of A-line muscle cells occurs through A5.2, A6.4, A7.8, and A8.16 of the 16-cell, 32-cell, 64-cell, and 112-cell stages, respectively (Fig. 6.1a, c). At the neural-plate stage, A9.31, a daughter of A8.16, becomes restricted to muscle fate, while the other daughter, A9.32, becomes part of the nervous system (Fig. 6.1a, c).

Multiple steps of A-line muscle fate induction involve sequential inputs from FGF, the TGFβ-family member Nodal, and the Delta signaling pathway in *Ciona* (Fig. 6.5). First, *FGF9/16/20*, expressed zygotically in vegetal cells from the 16-cell stage, likely activates, via MEK/ERK signaling, transcription of *Nodal* in b6.5 at the 32-cell stage (Fig. 6.5a, e). Nodal signals are required for the correct specification of all of the A7.8 derivatives, which will form the lateral part of the neural plate and the A-line muscle (Fig. 6.5b, f). Ablation of b5.3 (the precursor of b6.5) or b6.5 itself, and inhibition of Nodal signaling both lead to the loss of the entire lateral neural plate including the A-line muscle. The cells that had been destined to become muscle instead develop into the medial neural plate.

Next, at the 64-cell stage, Nodal acts, in part, via the induction of another ligand, Delta2, in a small group of lateral cells (A7.6, b7.9, and b7.10) that surround A7.8 (Fig. 6.5c, g). Following the division of A7.8 into A8.15 (neural fate) and A8.16 (neural and muscle fates), Delta2-expressing cells remain in contact with only A8.16 but not with A8.15 (Fig. 6.5c, g). Ectopic expression of Delta2 results in the specification of additional muscle cells within the neural plate; however, the additional muscle cells are derived only from A7.8 descendants, which receive Nodal signals at the 64-cell stage. Thus, it is likely that Nodal acts both directly and indirectly via Delta2 induction in A-line muscle specification.

The cell division of A8.16 takes place along the A–P axis, generating an anteriorly positioned A9.32 (fated to become neural cells) and its posteriorly positioned sister cell A9.31 (fated to become muscle cells). ERK1/2 activation is observed only in A9.31 (Fig. 6.5h). FGF9/16/20 is likely to be the ligand responsible for ERK1/2 activation, as *FGF9/16/20* is expressed throughout the A-line neural plate precursors at this stage. Inhibition of MEK (the kinase acting upstream of ERK1/2) from the early gastrula stage, or expression of a dominant-negative form of the FGF receptor in the A5.2 lineage, is sufficient to block the muscle fate specification of A9.31. The fate-altered A9.31 then

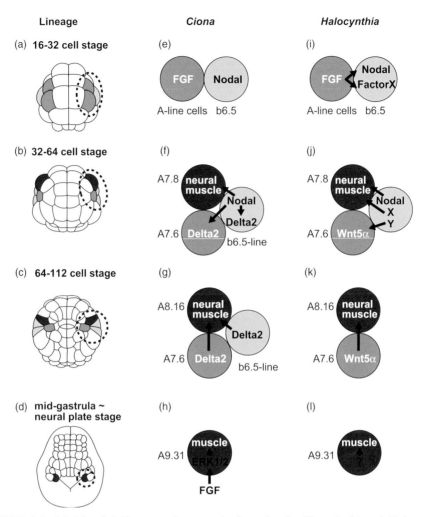

FIGURE 6.5 Models of A-line secondary muscle formation in *Ciona* (e–h) and *Halocynthia* (i–l). Blue circles, A-lineage cells; yellow circles, b-lineage cells; red circles, A-line secondary muscle lineage. Positions of these cells at the relevant developmental stages are indicated by dashed circles on the embryo schematics (a–d). Arrows indicate interactions between the cells and signaling molecules. Cells expressing *FGF9/16/20* are labeled with "FGF." For simplicity, the b-line cells (yellow) involved in these processes are represented as a single circle labeled "b6.5-line" (b7.9 and b7.10 at the 64-cell stage, and b8.20, b8.19, b8.18, and b8.17 at the 112-cell stage). (Modified from Hudson and Yasuo (2008) *Biol Cell*, **100**, 266–277.)

adopts the neural fate shared by its sister cell. Thus, it is likely that FGF/MEK/ERK signals act during the final step of muscle fate specification in the A-lineage.

In *Halocynthia*, however, the signaling pathways involved in A-line muscle induction appear to be considerably different from those in *Ciona* (Fig. 6.5). Similar to *Ciona*, in *Halocynthia*, FGF9/16/20-MEK signaling is required for the transcriptional activation of *Nodal* in b6.5 (Fig. 6.5i), and Nodal signals are required for the patterning of the A-line neural precursors along the medial–lateral axis. However, Nodal signals are not required for the specification of A-line muscle cells. Therefore, in *Halocynthia*, an unidentified signal, "X," which acts downstream of FGF9/16/20 in the b6.5 lineage, fulfills the role of *Ciona* Nodal during the induction of A-line muscle fate (Fig. 6.5i). In addition, the Notch signaling pathway does not appear to be required for secondary muscle development in *Halocynthia*. A *Halocynthia Delta* is expressed more broadly than *Ciona Delta2*, and it is not yet known whether the *Halocynthia Delta* is a target of Nodal signals (Fig. 6.5).

Despite the difference in Notch signaling requirements, A7.6 is necessary for A7.8-line muscle formation in *Halocynthia*. The A7.6 signal may involve *Wnt5α*, which is strongly expressed in A7.6 at the 64-cell stage (Fig. 6.5j). The Wnt5α expression in A7.6 depends on the b5.3 lineage, although neither Nodal nor factor X is involved. Therefore, another FGF9/16/20-independent signal, "Y," derived from the b6.5 lineage is required for *Wnt5α* expression in A7.6 (Fig. 6.5j). However, the role of Wnt5α during A-line muscle specification remains to be determined. Furthermore, in *Halocynthia*, FGF/MEK signals are not involved in the final division that generates A-line muscle precursors (Fig. 6.5l). Inhibition of FGF signal transduction in the entire A-lineage via injection of MO against the Ets transcription factor does not disrupt A-line muscle formation, and A8.16 can autonomously generate muscle fate when isolated late in its cell cycle. Further research may elucidate more precise signaling mechanisms in A-line muscle differentiation in *Ciona* and *Halocynthia*, and thereby highlight changes in the role of GRNs associated with evolutionarily related changes in their developmental blueprints.

6.2.2   b-Line: Lineage and Specification Mechanisms

Newly hatched *Halocynthia* and *Ciona* larvae are about 1.5 and 1 mm in length, respectively. Making a larger larva with a longer tail might have been accomplished by the creation of additional muscle cells either by expanding muscle developmental fate to additional cells, or by increasing the number of precursor cell divisions, or by both means. A comparison of *Ciona* and *Halocynthia* b-line muscle formation reveals that both routes were taken during evolution of the two species. The *Ciona* and *Halocynthia* larval tails develop 36 and 42 muscle cells, respectively. In *Ciona*, the b-line muscle precursors are b5.3, b6.5, and b7.9 of the 16-cell, 32-cell, and 64-cell embryo, respectively (Fig. 6.1a, d); b7.9 retains muscle fate and gives rise to two muscle cells in addition to epidermis, tail nerve cord, and endodermal strand. By contrast, in *Halocynthia*, both b7.9 and b7.10 of the 64-cell embryo, and b8.17 and b8.19 of the 112-cell embryo retain muscle fate; b8.17 produces three muscle cells, tail nerve cord, and endodermal strand, whereas b8.19 produces two muscle cells, epidermis, tail nerve cord, and trunk nerve cord cells (Fig. 6.1a, d).

What events lead to the differences observed between *Ciona* and *Halocynthia* larval tail muscles? One is the difference in the number of b-line precursor cell divisions. The b7.9 lineage produces four and six cells in *Ciona* and *Halocynthia*, respectively. This suggests that the cell cycle control of muscle precursors may be differentially regulated between these two species. Another reason for the disparities is that the position of the b7.10 lineage relative to muscle-inducing cells is slightly different between the two species, resulting in a situation whereby muscle-inducing cells are in contact with cells of both the b7.9 and b7.10 lineages in *Halocynthia*, but only with the b7.9 lineage in *Ciona* (if contact with the b7.10 lineage does occur, it is to a lesser extent than in *Halocynthia*). Indeed, in *Halocynthia*, no differences are reported in the mechanisms used by b7.9 and b7.10 lineages for muscle induction; both rely on FGF9/16/20 and Wnt5α, and become insensitive to inhibition of MEK signaling by the 130-cell stage. A differential response between competent cells to inductive cellular interactions can be achieved via a threshold in responsiveness to the signal, depending on the area of cell surface contact between inducing and responding cells. Further work is clearly needed in order to determine the precise embryological and molecular mechanisms involved in these evolutionary differences of the secondary muscle lineages of *Ciona* and *Halocynthia*. Once elucidated, the results will contribute to an understanding of the evolutionary alteration of GRNs.

SELECTED REFERENCES

BROWN, C., JOHNSON, D.S., SIDOW, A. (2007) Functional architecture and evolution of transcriptional elements that drive gene coexpression. *Science*, **317**, 1557–1560.

HUDSON, C., YASUO, H. (2005) Patterning across the ascidian neural plate by lateral Nodal signalling sources. *Development*, **132**, 1199–1210.

HUDSON, C., YASUO, H. (2006) A signalling relay involving Nodal and Delta ligands acts during secondary notochord induction in *Ciona* embryos. *Development*, **133**, 2855–2864.

HUDSON, C., LOTITO, S., YASUO, H. (2007) Sequential and combinatorial inputs from Nodal, Delta2/Notch and FGF/MEK/ERK signalling pathways establish a grid-like organisation of distinct cell identities in the ascidian neural plate. *Development*, **134**, 3527–3537.

HUDSON, C., YASUO, H. (2008) Similarity and diversity in mechanisms of muscle fate induction between ascidian species. *Biol Cell*, **100**, 266–277.

KIM, G.J., YAMADA, A., NISHIDA, H. (2000) An FGF signal from endoderm and localized factors in the posterior-vegetal egg cytoplasm pattern the mesodermal tissues in the ascidian embryo. *Development*, **127**, 2853–2862.

KUSAKABE, T., YOSHIDA, R., IKEDA, Y., TSUDA, M. (2004) Computational discovery of DNA motifs associated with cell type-specific gene expression in *Ciona*. *Dev Biol*, **276**, 563–580.

MARIKAWA, Y., YOSHIDA, S., SATOH, N. (1995) Muscle determinants in the ascidian egg are inactivated by UV irradiation and the inactivation is partially rescued by injection of maternal mRNAs. *Roux's Arch Dev Biol*, **204**, 180–186.

MEEDEL, T.H., LEE, J.J., WHITTAKER, J.R. (2002) Muscle development and lineage-specific expression of CiMDF, the MyoD-Family Gene of *Ciona intestinalis*. *Dev Biol*, **241**, 238–246.

NISHIDA, H. (1992) Regionality of egg cytoplasm that promotes muscle differentiation in embryo of the ascidian, *Halocynthia roretzi*. *Development*, **116**, 521–529.

NISHIDA, H., SAWADA, K. (2001) macho-1 encodes a localized mRNA in ascidian eggs that specifies muscle fate during embryogenesis. *Nature*, **409**, 724–729.

NISHIDA, H. (2005) Specification of embryonic axis and mosaic development in ascidians. *Dev Dyn*, **233**, 1177–1193.

NISHIDA, H. (2012) The maternal muscle determinant in the ascidian egg. *WIREs Dev Biol*, **1**, 425–433.

TOKUOKA, M., KUMANO, G., NISHIDA, H. (2007) FGF9/16/20 and Wnt-5α signals are involved in specification of secondary muscle fate in embryos of the ascidian, *Halocynthia roretzi*. *Dev Genes Evol*, **217**, 515–527.

WANG, W., CHRISTIAEN, L. (2012) Transcription enhancers in ascidian development. *Curr Top Dev Biol*, **98**, 147–172.

WHITTAKER, J.R. (1973) Segregation during ascidian embryogenesis of egg cytoplasmic information for tissue-specific enzyme development. *Proc Natl Acad Sci USA*, **70**, 2096–2100.

YAGI, K., TAKATORI, N., SATOU, Y., SATOH, N. (2005) *Ci-Tbx6b* and *Ci-Tbx6c* are key mediators of the maternal effect gene *Ci-macho1* in muscle cell differentiation in *Ciona intestinalis* embryos. *Dev Biol*, **282**, 535–549.

CHAPTER 7

ENDODERM

The endodermal tissue of the *Ciona* tadpole larva consists of ~500 cells and constitutes the trunk endoderm and tail endodermal strand (Fig. 7.1b). Ascidian embryogenesis is characterized by the development of larvae with no open mouth or no anus. Both stomata eventually open in the course of juvenile formation during metamorphosis. This type of precocious development appears to be an ascidian lifestyle strategy with the goal of quickly becoming a sessile juvenile. Accordingly, ascidian embryonic endodermal cells do not exhibit distinct functional differentiation, although in later developmental stages they differentiate into various endodermal tissues and organs of juveniles. However, even in earlier stages, the endodermal cells play pivotal roles in the establishment of the chordate body plan via signaling to induce the development of mesodermal and ectodermal tissues, as discussed in other chapters.

7.1 LINEAGE

The endoderm is originated from A-, B-, and b-line cells (Fig. 2.2, Fig. 2.3, and Fig. 7.1). The A-line potential to form endodermal tissues is segregated into A6.1 and A6.3 at the 32-cell stage, the former being restricted to give rise to endoderm. At the 64-cell stage, A7.5 (a daughter cell of A6.3) also becomes fate restricted to endoderm alone (Fig. 7.1a, b). By contrast, the B-line endodermal potential is inherited from B6.1 and B6.3 at the 32-cell stage, and then B7.1, B7.2, and B7.6 at the 64-cell stage. B7.1 is a precursor of endoderm in the tailbud embryo trunk, while B7.2 gives rise to ventral endoderm and endodermal strand (Fig. 7.1a, b). The endodermal strand cells are also derived from B7.6, which includes primordial germ cell precursors (see Chapter 14). The b-line endoderm is a descendent of b8.17, which also yields muscle and nerve cord progeny. b8.17 endodermal descendants occupy the caudal tip of the endodermal strand (Fig. 7.1b). Thus, with respect to its origin, the larval endoderm is subdivided into anterior and posterior parts that are derived from A- and B-line blastomeres, respectively (Fig. 7.1b). The anterior and posterior endoderm regions function independently by sending different signals to neighboring cells/tissues.

As previously noted, the larval endoderm develops into a variety of adult endodermal organs, including the esophagus, stomach, intestine, branchial sac, and endostyle (Fig. 7.1c, d). Lineage tracing of endodermal cells from the larval to juvenile stages reveals that the topographic position of each prospective region in the larval fate map corresponds to that of the adult organ, suggesting that marked rearrangement of endodermal cells does not occur during metamorphosis (Fig. 7.1c, d). However, the boundaries

Developmental Genomics of Ascidians, First Edition. Noriyuki Satoh.
© 2014 John Wiley & Sons, Inc. Published 2014 by John Wiley & Sons, Inc.

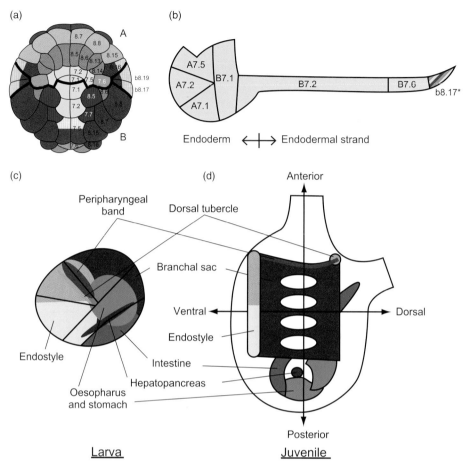

FIGURE 7.1   Fate map showing the relationship of larval endoderm to adult endodermal tissues. (a, b) Lineage of larval endoderm. (a) The 112-cell embryo showing blastomeres of endodermal lineage (orange yellow). (b) Lineage and clonal organization of endoderm and endodermal strand. The asterisk denotes the cells that give rise to tissues in addition to endoderm. (c, d) Relationship of larval to adult endodermal tissues. (c) Mid-saggital section of the trunk endodermal region of a larva showing the developmental fates of various endodermal regions. The anterior of the larva is to the left and dorsal is up. (d) Endodermal organs of a juvenile. The colors show the lineage relationships of larval to juvenile tissues (c, d: Adapted from Hirano and Nishida (2003) *Dev Genes Evol*, **210**, 55–63.)

between clones descended from early blastomeres do not correspond to the boundaries between adult endodermal organs, suggesting that fate specification of juvenile endodermal cells is likely a position-dependent rather than lineage-dependent process.

7.2   SPECIFICATION MECHANISMS

During the early developmental stages of many metazoans, including cnidarians, and sea urchins, maternally provided β-catenin plays a pivotal role in the specification of endodermal cells that are located in the vicinity of the vegetal pole. Deletion and transplantation experiments using *Halocynthia* egg cytoplasm have demonstrated that the presence of maternal factors are responsible for the autonomous differentiation of endodermal cells in the vegetal cytoplasm. In *Ciona*, after fertilization, egg cytoplasmic β-catenin is translocated only into nuclei of cells of the vegetal hemisphere (Fig. 7.2a, b). Suppressing the function of β-catenin leads to a dearth of AP activity, which characterizes ascidian endodermal cells. By contrast, the ectopic expression of β-catenin induces the differentiation of additional endodermal cells (Fig. 7.2c, d). The identity of vegetally localized factors that promote nuclear localization of β-catenin is not known yet and remains to be elucidated in various animals.

Maternal β-catenin activates the zygotic expression of genes involved in transcription (e.g., *Lhx-3*, *TTF-1* or *titf1*, *FoxA*, and *FoxD*), signaling (e.g., *FGF9/16/20*, *Eph*, *frizzled3/6*, and *lefty/antivin*), and other functions (e.g., *cadherinII*, *protocadherin*, and *netrin*). These factors play roles in the specification of cell fates in the vegetal hemisphere. The downstream genes *FGF9/16/20*, *frizzled3/6*, *lefty/antivin*, and *Eph* suggest that β-catenin regulates various signaling cascades, such as the FGF, Wnt, and Nodal pathways, which induce mesodermal development (see Chapter 11).

Of the β-catenin downstream regulatory genes, *FoxA* is expressed in the vegetal blastomeres at the 16-cell stage and a LIM-homeobox gene, *Lhx-3b*, is expressed in A- and B-line endodermal cells at the 32-cell stage. Functional suppression and ectopic expression experiments show that *Lhx-3b* is necessary and sufficient for endodermal differentiation as assessed by the differentiation marker, AP. The expression of *TTF-1* commences at the 64–76-cell stage, and is also sufficient to induce AP expression. Therefore, a prominent signaling pathway leading to autonomous endodermal differentiation is maternal β-catenin → FoxA → Lhx3 and/or TTF-1 → specific structural genes.

However, β-catenin enters the nuclei not only of presumptive endodermal cells, but also of other vegetal cells that are not fated to become endoderm. Furthermore, the expression of β-catenin downstream genes such as *FGF9/16/20* and *FoxA* are not exclusive to the endodermal lineage, indicating that the function of β-catenin is not restricted to the endoderm lineage. Therefore, the embryo must have additional mechanisms that ensure proper segregation of the endodermal lineage. As shown in Fig. 7.3A, in *Halocynthia* embryos, three pairs of vegetal cells at the 16-cell stage (A5.1, A5.2, and B5.1) are mesendoderm cells. These cells divide asymmetrically to give rise to mesoderm (A6.2, A6.4, and B6.2) in the marginal zone and endoderm (A6.1, A6.3, and B6.1) in the vegetal pole region. After cell division, the mesodermal cells begin to express *ZicN*, the *Halocynthia* ortholog of *Ciona ZicL*. The endodermal cells express *Lhx-3b* that is sufficient and required for endodermal fate and suppresses mesodermal fate. On the other hand, *ZicN* is sufficient and required for mesodermal fate and represses endodermal fate. In other words, the separation of the expression of these two transcription factors defines the mesoderm and endoderm fates.

Furthermore, the mechanism that segregates mesendoderm into mesoderm and endoderm has recently been discovered in *Halocynthia* embryos. These fates are segregated by partitioning of the asymmetrically localized *Not* mRNA from the mesendoderm cell to its mesodermal daughter (Fig. 7.3B). This is accomplished by (i) migration of the mesendoderm cell nucleus that is transcribing *Not* mRNA to the future mesoderm-forming region (Fig. 7.3c), (ii) release of *Not* mRNA from the nucleus (Fig. 7.3d), (iii) Wnt5α-dependent local retention of the mRNA (Fig. 7.3e), and (iv) repositioning of the mitotic spindle to the center of the cell (Fig. 7.3e) and cell division (Fig. 7.3f). Notably, the nuclear

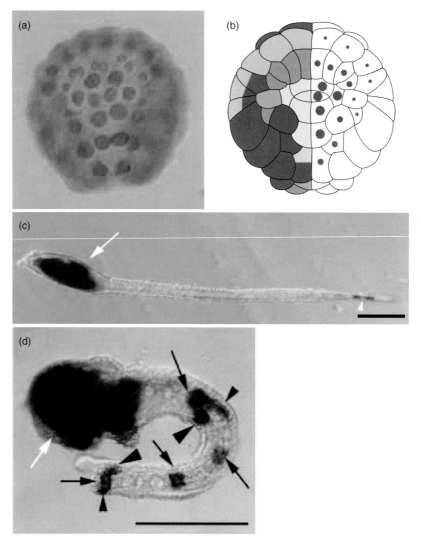

FIGURE 7.2   Nuclear β-catenin accumulation in *Ciona* embryos and role of endoderm cell specification. (a, b) A 112-cell embryo showing elevated levels of nuclear β-catenin in vegetal blastomeres. (a) Antibody staining and (b) a schematic drawing showing the fates of blastomeres that accumulate β-catenin in their nuclei. The size of the nucleus corresponds to the relative quantity of nuclear β-catenin accumulation. Yellow, endoderm; pink, notochord; green, mesenchyme; light green, trunk lateral cells; red, muscle; and magenta, nerve cord. (c) Histochemical staining of alkaline phosphatase (AP) with a mock-electroporated larva (control). AP activity is found in endoderm (arrow) and a few notochord cells of the tip of the tail (arrowhead). (d) A larva is electroporated with *Ci-fkh*(promoter) : *Ci-β-catenin*. In addition to expression of AP in endoderm cells (white arrow), ectopic AP expression is evident in notochord (arrows), endodermal strand (large arrowheads), and nerve cord (small arrowheads). Bars, 100 μm. (Adapted from Imai *et al.* (2000) *Development*, **127**, 3009–3020.)

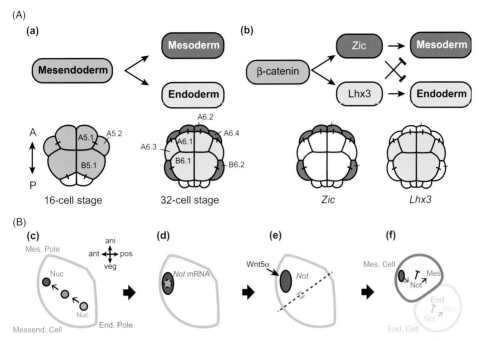

FIGURE 7.3 *Not* mRNA partitioning and mesendodermal fate segregation. (A) Schematic representation of fate segregation from the 16- to 32-cell stage in *Halocynthia* embryos. (a) Mesendodermal cells (orange) give rise to mesoderm (green) and endoderm (yellow). Bars that connect the
cells indicate that they are sister cells. (b) *Zic* (*ZicN*), expressed in mesodermal cells, and *Lhx3*,
expressed in endodermal cells, mutually suppress the development of the other fate. Cells expressing *Zic* and *Lhx3* are shown in green and yellow, respectively. (B) *Not* mRNA partitioning. (c) *Not*
mRNA (blue) is transcribed in the nucleus (green) as it migrates to the future mesoderm-forming
side of the mesendodermal cell. (d) Then, *Not* mRNA is delivered from the nucleus to the future
mesodermal cell cytoplasm at the transition from interphase to M-phase. (e) A Wnt5α-dependent
mechanism retains *Not* mRNA to the mesodermal pole as the mitotic spindle repositions to the
center of the cell. A dotted line indicates the plane of future cell division. (f) Subsequent cell
division partitions *Not* mRNA to the mesodermal precursor cell. *Not* mRNA translated in the
mesodermal cell inhibits the endodermal fate and promotes the mesodermal fate. The absence of
Not in endoderm permits endoderm differentiation. Nuc, nucleus; ani, animal; veg, vegetal; ant,
anterior; pos, posterior; Mesend, mesendoderm; Mes, mesoderm; End, endoderm. (Adapted from
Takatori *et al.* (2010) *Dev Cell*, **19**, 589–598.)

migration plays an unexpected role in the asymmetric cell divisions that segregate germ
layer fates in chordate embryos.

In summary, ascidian endoderm is specified through the combined effect of two mechanisms. The β-catenin dependent mechanism instructs the vegetal hemisphere cells to
become endoderm. When Not function is added on to that of β-catenin, the cell becomes
mesoderm. The cells that do not translate Not will proceed as endoderm.

7.3   DIFFERENTIATION MECHANISMS

The mechanisms underlying endoderm-specific gene expression remain to be elucidated by future studies. However, as discussed in Chapter 3, one advantage of the *Ciona* system is the vast collection of gene expression profiles developed for this organism. For example, WISH studies of 1013 randomly selected genes expressed in the tadpole larva demonstrate that 108 of them are expressed specifically in endodermal cells. Many endoderm-specific genes encode protein synthesis cofactors, tRNA synthetase, and ribosomal proteins. In addition, several genes are differentially expressed within the endodermal tissues. For example, a gene with no significant matches in the database is expressed only in the trunk endoderm (Fig. 3.4f). A gene that encodes a protein similar to mammalian SOUL is expressed in the posterior endoderm and endodermal strand (Fig. 3.4g). Another gene is expressed only in a portion of the endodermal strand. This heterogeneity of the endoderm-specific gene expression may be related to the developmental fates of the endodermal cells after metamorphosis. These endoderm-specific genes may be useful tools for investigating the developmental mechanisms that govern endodermal differentiation, for example, in relation to the regulatory cascade downstream of *Lhx-3* and *TTF-1*.

SELECTED REFERENCES

HIRANO, T., NISHIDA, H. (2003) Developmental fates of larval tissues after metamorphosis in the ascidian, *Halocynthia roretzi*. II. Origin of endodermal tissues of the juvenile. *Dev Genes Evol*, **210**, 55–63.

IMAI, K., TAKADA, N., SATOH, N., SATOU, Y. (2000) β-catenin mediates the specification of endoderm cells in ascidian embryos. *Development*, **127**, 3009–3020.

NISHIDA, H. (2005) Specification of embryonic axis and mosaic development in ascidians. *Dev Dyn*, **233**, 1177–1193.

RISTORATORE, F., SPAGNUOLO, A., ANIELLO, F., BRANNO, M. *et al*. (1999) Expression and functional analysis of *Cititf1*, an ascidian NK-2 class gene, suggest its role in endoderm development. *Development*, **126**, 5149–5159.

SATOU, Y., IMAI, K.S., SATOH, N. (2001) Early embryonic expression of a LIM-homeobox gene *Cs-lhx3* is downstream of beta-catenin and responsible for the endoderm differentiation in *Ciona savignyi* embryos. *Development*, **128**, 3559–3570.

TAKATORI, N., KUMANO, G., SAIGA, H., NISHIDA, H. (2010) Segregation of germ layer fates by nuclear migration-dependent localization of *Not* mRNA. *Dev Cell*, **19**, 589–598.

CHAPTER 8

EPIDERMIS

The outer surface of an ascidian larva is composed of a layer of epidermal cells, and is enclosed by tunic. The larval tunic is a double-layered structure. An inner tunic encloses the entire larval body while an outer tunic surrounds only the tail region (cf., Fig. 3.4a, b). The former gives rise to the juvenile and adult tunic, whereas the latter is discarded during the course of tail regression at metamorphosis.

The ascidian embryonic epidermis has at least two unique features with respect to metazoan embryogenesis. The first is early fate determination, a general feature of ascidian development. As early as the 16-cell stage, the b5.4 pair is restricted to generate only the epidermis, and, through seven synchronous divisions, they yield 256 epidermal cells in the larval tail (see below). The second attribute reflects the fact that tunicates, including ascidians, are the only animal group that can independently synthesize cellulose, a process that occurs in the larval epidermis. Accordingly, *Ciona* provides an outstanding experimental system in which to explore the molecular mechanisms of cellulose biosynthesis.

8.1 LINEAGE AND SPECIFICATION MECHANISMS

8.1.1 Lineage

The epidermis is derived entirely from cells of the animal hemisphere of early embryos, and all four pairs of a5.3, a5.4, b5.3, and b5.4 of the 16-cell embryo contribute to the epidermal lineage (Fig. 2.2 and Fig. 2.3). As previously discussed, the b5.4 pair is restricted to generate only the epidermis at the 16-cell stage, and through its progeny—including b6.7 and b6.8 of the 32-cell embryo, as well as b7.13, b7.14, b7.15, and b7.16 of the 64-cell embryo—the b5.4 pair gives rise to 256 epidermal cells in the ventral part of the mid-tailbud embryo tail (Fig. 8.1c). Specifically, each of the b7.13–b7.16 pairs gives rise to a clone of 64 epidermal cells (Fig. 8.1c). Meanwhile, the a6.6, a6.8, and b6.6 pairs become epidermis-restricted at the 32-cell stage. The first two pairs give rise to a total of 256 epidermal cells in the lateral and ventral trunk regions of the mid-tailbud embryo (Fig. 8.1c). The b6.6 pair forms 128 epidermal cells in the lateral part of the tail (Fig. 8.1c). At the 64-cell stage, the a7.14 pair (a5.4-line) also becomes restricted to give rise to 64 epidermal cells in the dorsal trunk region (Fig. 8.1c). Additional epidermal cells of the dorsal trunk region are derived from the b8.18 and b8.20 pairs (each forms 32 epidermal cells), the anterior trunk region arises from a8.18 and a8.20 pairs, and the anterior dorsal trunk region epidermis arises from the a8.26 pair (Fig. 8.1c). Therefore, the total number of epidermal cells of the mid-tailbud embryo is estimated to be 800.

Developmental Genomics of Ascidians, First Edition. Noriyuki Satoh.
© 2014 John Wiley & Sons, Inc. Published 2014 by John Wiley & Sons, Inc.

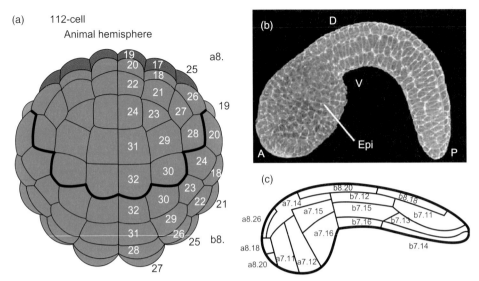

FIGURE 8.1   Diagram illustrating the lineage and clonal organization of ascidian embryo epidermal cells. (a) The 112-cell embryo showing blastomeres of epidermal lineage (green). Animal pole view. (b) 3-D image of *Ciona* tailbud embryo epidermis (Epi). A, anterior; P, posterior; D, dorsal; V, ventral. (Adapted from Nakamura *et al.* (2012) *Dev Biol*, **372**, 274–284.) (c) The clonal organization of the epidermal cells of a tailbud embryo, with domains of cells derived from epidermis-restricted blastomeres at the 76- and 110-cell stages. The domains derived from a8.18 and a8.20 are developed into larval papillae.

This number is maintained until the larval stage. As shown in Fig. 8.1, certain epidermal progenitor cell clones cover the entire embryo by the epibody stage without changing their positions relative to neighboring clones.

8.1.2   Specification Mechanisms

8.1.2.1   *Maternal Factor(s)*   Experiments involving the deletion or transplantation of *Halocynthia* egg cytoplasm indicate that maternally provided factor(s), which are distributed rather broadly in the animal hemisphere, are responsible for the autonomous specification of epidermal cells. Compared to the molecular cascades describing mesoendoderm specification, less is known about the molecular events that specify the ectoderm of early ascidian embryos. The maternal transcription factor GATAa (GATA4/5/6) is initially distributed throughout the embryo. Its gene activity is progressively repressed in the vegetal pole region via accumulating maternal β-catenin (Fig. 8.2). Once restricted to the animal hemisphere, GATAa directly activates two zygotic ectodermal genes, *Fog* (*friend of GATA*) and *Otx* (Fig. 8.2). *Fog* is expressed from the 8-cell stage throughout the ectoderm, while *Otx* is turned on from the 32-cell stage in neural precursors only (Fig. 8.2). Whereas the enhancers of both genes contain critical and interchangeable GATA sites, their distinct activation patterns stem from the additional presence of two Ets binding sites in the *Otx* enhancer. Initially characterized as activating elements in the neural lineages, these Ets sites additionally act as repressors in nonneural lineages, and restrict GATA-mediated activation of *Otx*.

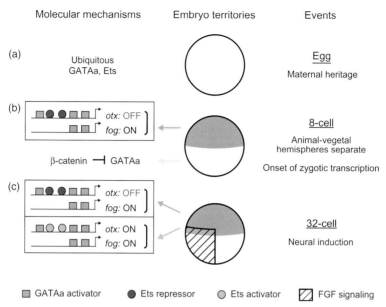

FIGURE 8.2   Model for early animal patterning in ascidians. Schematic views of the different events taking place in the (a) early *Ciona* egg, (b) 8-cell, and (c) 32-cell embryo and the corresponding molecular mechanisms defining the ectodermal territories. (Adapted from Rothbächer *et al.* (2007) *Development*, **134**, 4023–4032.)

8.1.2.2   Cell Interactions and Regionalization

The tail epidermis of *Ciona* tailbud embryos is divided into the midline, medio-lateral, and lateral regions (Fig. 8.3). As shown in Fig. 8.3a–c, the dorsal midline of the tail epidermis originates from b8.18 and b8.20, whereas the lateral region derives from b6.6. Peripheral neurons develop only in the dorsal and ventral midline epidermis (Fig. 8.3i).

When whole posterior animal blastomeres (a6.6–b6.8) are isolated and cultured at the 32-cell stage, the explants give rise to only lateral tail epidermis, suggesting that the signals originating from the vegetal hemisphere induce the midline and medio-lateral fates. The signal is FGF9/16/20, which is expressed in vegetal cells at the 32-cell stage. Nodal is likely to act downstream of FGF9/16/20 during the induction of the dorsal-midline tail epidermis derived from b7.9, b7.10, and b7.11 (Fig. 8.3e). However, as Nodal and Delta2 are expressed in all b7.9 and b7.10 progeny, these ligands do not explain why a single daughter of each of these precursors adopts a midline epidermal fate. In addition, the precise timing of the induction and a potential role for Delta2 in the process remains to be determined.

The ventral midline epidermis originates from b8.27, b8.28, b8.31, and b8.32 (Fig. 8.3a–c). Among these cells, b8.27 is the only one that exclusively gives rise to midline epidermis, whereas the other cells are bipotential midline/medio-lateral epidermis precursors (Fig. 8.3a–c). The two fates segregate during the next, medio-lateral, cell division. Ventral midline induction involves bone morphogenetic protein (BMP) signaling. Gain-of-function experiments indicate that activating the BMP pathway during gastrulation is sufficient to convert most b-line epidermal cells to the ventral midline fate. Conversely, functional suppression of antidorsalizing morphogenetic protein (ADMP) leads to a specific loss of the ventral, but not dorsal, epidermis

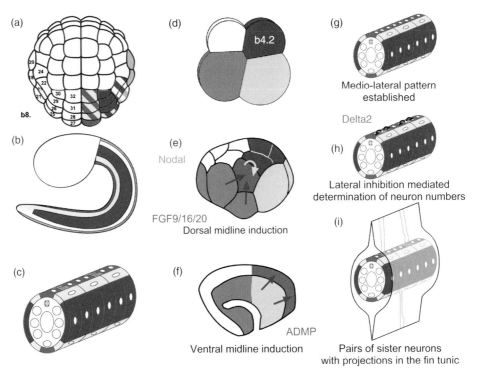

FIGURE 8.3   Tail epidermis cell lineages (a–c) and their regional specification mechanisms (d–i) in *Ciona* embryos. (a) Color-coding illustrates the clonal basis for defining dorso-ventral regions in the tail epidermis. Pink, dorsal midline domain; light yellow, dorso-lateral domain; blue, lateral domain; yellow, ventro-lateral domain; purple, ventral midline. (b, c) Schematic representation of tail medio-lateral patterning. (d) The tail epidermis derives from the posterior animal b4.2 blastomeres. (e) At the 32-cell stage, FGF9/16/20 induces *Nodal* and dorsal midline identity in the b6.5 blastomere (red). (f) At the mid-gastrula stages, ADMP induces ventral midline fate in the overlying epidermis. (g) These first inductive processes lead to a medio-laterally patterned epidermis comprising eight rows of cells in cross section. (h, i) Finally, Notch signaling controls the number of caudal epidermal sensory neurons in the midline through lateral inhibition. (Adapted from Pasini *et al.* (2006) *PLoS Biol*, **4**, e225.)

midline (Fig. 8.3f). Consistent with a direct inducing role, ADMP is expressed at the mid-gastrula stage in the B-line ventral midline vegetal cells, in direct contact with epidermal midline precursors (Fig. 8.3f).

In *Halocynthia* tailbud embryos, the expression profiles of eight epidermis-specific genes define six distinct areas of the epidermis. It is intriguing to think about how these region-specific expression patterns in both tail epidermis and trunk epidermis, are controlled in relation to the above-mentioned molecular cascade involved in epidermal compartmentalization.

8.2   ADHESIVE ORGAN

The adhesive organ of *Ciona* larvae, which consists of three cone-shaped protrusions called palps or papillae, develops in the anterior-most part of the tadpole larva and serves to attach the larva to a solid surface at the time of metamorphosis. Lineage studies show that one of the daughter cells of the left and right a8.18 forms the dorsal papillae, whereas a8.20 and a8.20 give rise to the left and right ventral papillae, respectively (Fig. 2.3).

The development of papillae from a-line cells requires interactions with A-line cells. As previously discussed, FGF signals emanating from the vegetal blastomeres, acting together with maternally-expressed GATA and Ets, lead to the expression of *Otx* in the sensory vesicle. With respect to left–right asymmetry (Chapter 10), in *Halocynthia*, the single copy gene, *Hr-Pitx*, is expressed in the left epidermis at the tailbud stage. In addition, *Hr-Pitx* is expressed in the papilla-forming region from the neurula to tailbud stages. Studies of transcriptional regulatory modules for the papilla-specific expression of *Pitx* identified Otx and Fox binding sites in 850–1211-bp upstream of the *Pitx* translational start site. morpholino oligonucleotides (MO)-mediated reductions in *Otx* expression result in a loss of *Pitx* expression in the papilla-forming region. A similar genetic cascade is active during the development of the *Xenopus* adhesive organ, suggesting a conserved genetic mechanism for development of adhesive organs between ascidians and vertebrates.

8.3   CELLULOSE BIOSYNTHESIS

Cellulose is a hydrogen-bonded β-1,4-linked glucan microfibril that constitutes the largest biomass on Earth. Cellulose is produced by a taxonomically diverse group of organisms, including plants, algae, bacteria, protists, and fungi. Cellulose is synthesized by large multimeric protein complex. The only known components of the complex are the cellulose synthase (*CesA*) and cellulase proteins. Owing to the complexity of the component proteins, the molecular mechanisms involved in cellulose biosynthesis are not yet clear. For example, the *Arabidopsis* genome contains 10 *CesA* genes (*CesA1-10*) that encode proteins with homology to bacterial cellulose synthases. *CesA1, CesA3, CesA6, CesA2, CesA5,* and *CesA9* have been shown to associate with the CesA complexes active during primary wall formation, while *CesA4, CesA7,* and *CesA8* have been reported to be part of the CesA complexes responsible for secondary wall cellulose biosynthesis. However, the exact number and the stoichiometry of CesA proteins involved in primary and secondary wall formation are still unknown.

Urochordates comprise the only animal population that is able to produce cellulose independently. The genomes of *Ciona intestinalis* and *Ciona savignyi* each contain a single copy of *CesA* (Fig. 8.4a–c). The *Ciona CesA* is likely to have originated through lateral transfer of a bacterial cellulose synthase gene into a urochordate ancestor. Interestingly, *Ci-CesA* encodes CesA protein in the N-terminal half and cellulase protein in the C terminal half, suggesting that a sophisticated machinery of cellulose biosynthesis (Fig. 8.4a–c). The gene is expressed in epidermal cells of tailbud embryos and larvae (Fig. 8.4d).

The function of *Ciona CesA* is clearly shown by analyses of a transposon-mediated insertional mutant, *swimming juveniles* (*sjs*). In this mutant, a tandem array of *Minos* is inserted 327–328 bp upstream of the *Ci-CesA* transcriptional start site (Fig. 8.4a). The *sj* mutant larvae initiate metamorphosis events in the trunk without undergoing tail resorption (Fig. 8.4e), and the mutant shows a drastic reduction of cellulose in the tunic (Fig. 8.4g). Functional suppression with MO of *Ci-CesA* leads to a phenotype similar to that of *sj*. These results indicate that *Ci-CesA* is responsible for cellulose biosynthesis, and

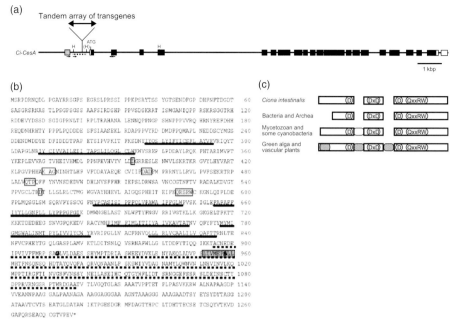

FIGURE 8.4 Structure and function of *Ci-CesA*, a gene encoding cellulose synthase. (a) Schematic representation of the gene displays the total 14-kb region. *Ci-CesA* contains 21 exons. Location of the *Minos* insertion site in the *sj* mutant genome is shown. Open and filled boxes represent *Ci-CesA* exons corresponding to the UTR and ORF, respectively. A gray box indicates the position of an EST. "H" labels the restriction sites of *HindIII*. (b) Ci-CesA protein structure. Numerals denote the number of amino acids from the putative translation-initiating methionine. Lines denote the positions of the transmembrane helices. The yellow and pink boxed characters represent the common motifs of uridine diphosphate (UDP)-dependent, polymerizing GT-2s (D, DxD, D, QxxRW) and the signature sequence of GH-6 proteins ([LIVMYA]–[LIVA]–[LIVT]–[LIV]–E–P–D–[SAL]–[LI]–[PSAG]), respectively. CesA belongs to GT-2 and cellulase to GH-6. The boxed characters represent the KAG and QTP motifs. The dotted line denotes the region demonstrating similarity to a GH-6 conserved domain. Bold characters represent the positions corresponding to the probable aspartic acid residues conserved in the catalytic core of GH-6s. (c) Comparison of the domain structure of cellulose synthases of various origins. Multiple alignments are shown in schematic drawings indicating the positions of the common motifs, depicted as D, DxD, D, and QxxRW. Only the cellulose synthase domain of Ci-CesA is represented. In comparison with ascidian and prokaryotic sequences, algal and plant sequences conserved an N-terminal zinc-binding domain and two insertions, the so-called plant-specific conserved region (CR-P) and a class-specific region, depicted in gray. (Adapted from Nakashima *et al.* (2004) *Dev Genes Evol*, **214**, 81–88.) (d–g) Assessing *CesA* function in ascidians through the use of the transposon-based mutant, *swimming juvenile* (*sj*). (d, e) *Ci-CesA* expression is not detected in the *sj* (e), but is evident in the trunk of a normal (wild-type) larva at the same age (d) (normal; Bar, 100 μm). (f, g) Cellulose production is lost in the *sj* (g), as compared to a normal larva of the same age (normal) (f). (Adapted from Sasakura *et al.* (2005) *Proc Natl Acad Sci USA*, **102**, 15134–15139.)

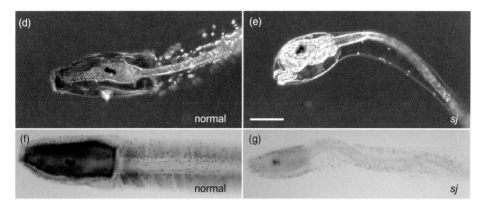

FIGURE 8.4   (*Continued*)

also coordinates metamorphosis events in the trunk and tail. Immunoprecipitation experiments using an anti-Ci-CesA antibody are now being conducted to identify components of the multimeric protein complexes.

The native form of cellulose is a fibrillar composite of two crystalline phases, the triclinic I_α and monoclinic I_β allomorphs. Allomorph ratios are species-specific, and give rise to natural structural variations in cellulose crystals. However, the mechanisms contributing to crystal formation are unknown. The two crystalline phases of cellulose are tailored to yield distinct structures during different developmental stages of the larvacean *Oikopleura dioica*. In *O. dioica*, larval cellulose consisting of the I_α allomorph constitutes the body cuticle fin, whereas adult cellulose consisting of the I_β allomorph frames a mucous filter-feeding device, the "house." As mentioned above, the *Ciona* genome contains a single copy of *CesA*, while the *Oikopleura* genome contains two copies: *Od-CesA1* and *Od-CesA2*. These genes are expressed in a mutually exclusive pattern, *Od-CesA1* in the body cuticle and *Od-CesA2* in the house, and each is responsible for producing different cellulose structures. Thus, *O. dioica* provides a unique model system in which to explore the possible connections between structural variations of cellulose crystalline phases and the underlying evolutionary genetics of cellulose biosynthesis.

SELECTED REFERENCES

BERTRAND, V., HUDSON, C., CAILLOL, D., POPOVICI, C. *et al.* (2003) Neural tissues in ascidian embryos is induced by FGF9/16/20, acting via a combination of maternal GATA and Ets transcription factors. *Cell*, **115**, 615–27.

NAKAMURA, M.J., TERAI, J., OKUBO, R., HOTTA, K. *et al.* (2012) Three-dimensional anatomy of the Ciona intestinalis tailbud embryo at single-cell resolution. *Dev Biol*, **372**, 274–284.

NAKASHIMA, K., YAMADA, L., SATOU, Y., AZUMA, J. *et al.* (2004) The evolutionary origin of animal cellulose synthase. *Dev Genes Evol*, **214**, 81–88.

NAKASHIMA, K., NISHINO, A., HORIKAWA, Y., HIROSE, E. *et al.* (2011) The crystalline phase of cellulose changes under developmental control in a marine chordate. *Cell Mol Life Sci*, **68**, 1623–1631.

PASINI, A., AMIEL, A., ROTHBÄCHER U., ROURE, A. *et al.* (2006) Formation of the ascidian epidermal sensory neurons: insights into the origin of the chordate peripheral nervous system. *PLoS Biol*, **4**, e225.

ROTHBÄCHER, U., BERTRAND, V., LAMY, C., LEMAIRE, P. (2007) A combinatorial code of maternal GATA, Ets and β-catenin-TCF transcription factors specifies and patterns the early ascidian ectoderm. *Development*, **134**, 4023–4032.

SASAKURA, Y., NAKASHIMA, K., AWAZU, S., MATSUOKA, T. *et al.* (2005) Transposon-mediated insertional mutagenesis revealed the functions of animal cellulose synthase in the ascidian *Ciona intestinalis*. *Proc Natl Acad USA*, **102**, 15134–15139.

YOSHIDA, K., UENO, M., NIWANO, T., SAIGA, H. (2012) Transcription regulatory mechanism of Pitx in the papilla-forming region in the ascidian, *Halocynthia roretzi*, implies conserved involvement of Otx as the upstream gene in the adhesive organ development of chordates. *Dev Growth Differ*, **54**, 649–659.

CHAPTER 9

NOTOCHORD

The notochord is the most prominent feature of chordates and has two distinct functions during early development. First, it is an axial organ of chordate embryos and acts as a source of inductive signals for neural tube patterning and paraxial mesoderm. Second, it is a supportive organ of the chordate larval tail. The elucidation of the genetic regulatory network that governs notochord development in ascidian embryos therefore guides our understanding of chordate evolution itself.

9.1 NOTOCHORD LINEAGE, CELL-FATE DETERMINATION, AND MORPHOGENESIS

The ascidian notochord is composed of a monolayer sheet of exactly 40 cells that rearrange to make a single rod of coin-shaped cells (Fig. 9.1). This number is invariant among known ascidian species, suggesting that the number of notochord precursor-cell divisions is controlled under tight evolutionary constraints.

As shown in Fig. 9.1A, the lineage of the 40 notochord cells is completely documented. The 40 cells are derived from 32 A-line and eight B-line blastomeres. The A-line notochord potential is inherited by A5.1 and A5.2 of the 16-cell embryo (Fig. 9.1b), A6.2 and A6.4 of the 32-cell embryo (Fig. 9.1c), and A7.3 and A7.7 of the 64-cell embryo (Fig. 9.1d). A7.3 and A7.7 become restricted to produce notochord and divide three times to form 32 notochord cells at the anterior and middle part of the tail (Fig. 9.1e, f). In addition, B5.1, B6.2, B7.3, and B8.6 of the 16-cell (Fig. 9.1b), 32-cell (Fig. 9.1c), 64-cell (Fig. 9.1d), and 112-cell embryo (Fig. 9.1e), respectively, are the presumptive notochord cells. The B8.6 pair becomes restricted to notochord, giving rise to eight notochord cells in the posterior part of the tail after two divisions (Fig. 9.1f). Therefore, every notochord cell has a history of nine divisions from the zygote stage to the ultimate differentiation. The A- and B-line notochord cells are sometimes called "primary" and "secondary" notochord cells, because of their different specification mechanisms.

The notochord is formed in the middle of the larval tail via a series of dynamic morphogenetic movements that include cell motility, changes in cell shape, cell rearrangements, and tissue deformation (Fig. 9.1B). In *Ciona*, at the late gastrula stage, 10 presumptive notochord cells form a semicircular arc around the anterior lip of the gastrula blastopore (Fig. 9.1g), and, in 75 min, undergo two more rounds of division along the A–P axis of the embryo to generate 40 notochord cells (Fig. 9.1h, i). At the end of the final divisions, the notochord cells form a monolayered epithelium (Fig. 9.1i). The next steps convert the monolayer of epithelial-like cells to a cylindrical rod, a single

Developmental Genomics of Ascidians, First Edition. Noriyuki Satoh.
© 2014 John Wiley & Sons, Inc. Published 2014 by John Wiley & Sons, Inc.

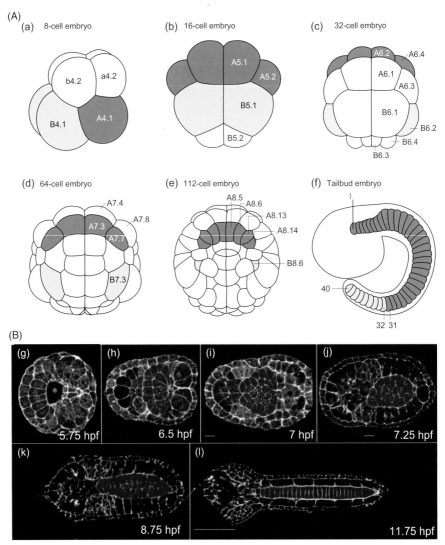

FIGURE 9.1 Development of the ascidian notochord. (A) Cell lineage. The notochord is derived from 32 A-line and 8 B-line cells. (a) The 8-cell embryo, (b) the 16-cell embryo, (c) the 32-cell embryo, (d) the 64-cell embryo, (e) the 112-cell embryo, and (f) the tailbud embryo. A7.3, A7.7, and B8.6 become restricted to give rise to the notochord. Every notochord cell has a history of nine divisions between the zygote stage and ultimate differentiation. (B) Ascidian notochord development from the onset of gastrulation to the completion of convergent extension. (g) The gastrula, and * indicates the blastopore. (h–j) The neurula. Two rounds of cell division generate 20 (h) and finally 40 notochord cells that form a monolayer epithelium (i). (j–l) Infolding (not shown here) and convergent extension transform the notochord precursor into a column of 40 stacked cells. *Ciona savignyi* embryos are stained with bodipy-phalloidin and imaged with a laser scanning confocal microscope. hpf, hours postfertilization. All images are dorsal view, with anterior to the left. (B: Adapted from Jiang and Smith (2007) *Dev Dyn*, **236**, 1748–1757.)

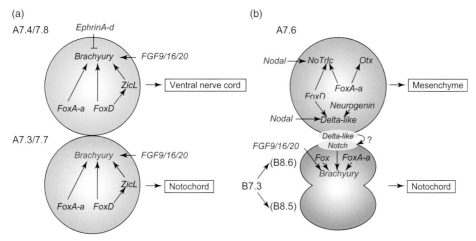

FIGURE 9.2 Regulatory interactions specifying notochord fates and *Brachyury* expression in notochord cells. (a) Key regulatory interactions generating *Ci-Bra* expression in A-line notochord cells (A7.3/A7.7) at the 64-cell stage. The expression of *Ci-Bra* in A7.4/7.8, nerve-cord precursors, is blocked by Eph signaling from neighboring animal cells. (Modified from Satou *et al.* (2009) *Biochim Biophys Acta*, **1789**, 268–273). (b) Key regulatory interactions generating *Ci-Bra* expression in B-line notochord cells (B7.3) at the 64-cell stage. Here, Delta-like and Notch/Delta signaling are involved in *Ci-Bra* expression. (Modified from Imai *et al.* (2006) *Science*, **312**, 1183–1187.)

cell in diameter. Medial intercalation, convergence, and extension of the individual cells produce an approximately threefold elongation of the notochord (Fig. 9.1j, k). Then, the second phase of elongation occurs as the notochord narrows medially and increases in volume (Fig. 9.1l). The mechanism by which the notochord volume increases differs among ascidian species. Some ascidians, including *Ciona*, produce extracellular pockets that will eventually coalesce to form a lumen running along the length of the notochord (see Section 9.3), and this postconvergent extension brings about an additional threefold elongation of the tail. The resulting notochord serves as a hydrostatic skeleton allowing for the locomotion of the swimming larva.

9.2   SPECIFICATION MECHANISMS AND THE GENE REGULATORY NETWORK

9.2.1   *Brachyury*: A Key Specifier Gene

A member of the T-box family, *Brachyury*, is a key regulatory gene for ascidian notochord formation. It is exclusively expressed in the notochord-fated A7.3, A7.7, and B8.6, and their descendants (Fig. 9.3a). The functional suppression of *Bra* blocks notochord development whereas the ectopic expression of *Bra* in endoderm cells redirects their fate to become notochord cells. The upstream and downstream signaling cascades of *Ci-Bra* are well characterized (Fig 9.2 and Fig. 9.3).

9.2.2   A-Line Genetic Cascade Upstream of *Bra*

Two genetic cascades are responsible for the exclusive expression of *Bra* in the primordial notochord cells. One is a cell-autonomous cascade, beginning with the actions of

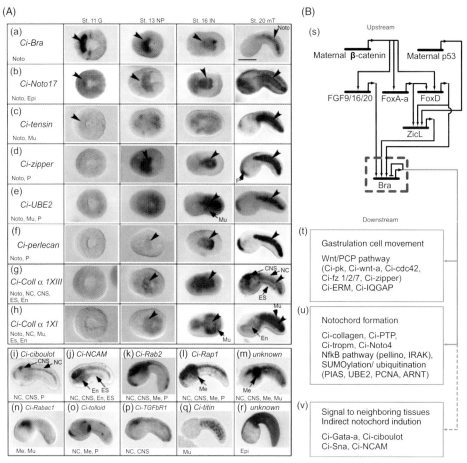

FIGURE 9.3   Genes involved in notochord formation. (**A**) Whole-mount *in situ* hybridization showing examples of organ- and tissue-specific expression profiles of *Ci-Bra* (a) and *Ci-Bra* downstream genes (b–r) in *C. intestinalis* tailbud embryos. The organs and tissues where each gene is specifically expressed are shown in the bottom left-hand corner. Noto, notochord; P, papilla; CNS, central nervous system; Me, mesenchyme; En, endoderm; ES, endodermal strand; NC, nerve cord; Mu, muscle; Epi, epidermis; TT, tail tip. Scale bar: 50 μm. Abbreviations along the top of the columns indicate the developmental stage. G, gastrula; NP, neural plate; lN, late neurula stage; mT, mid-tailbud stage. Arrowheads indicate the expression of notochord cells/notochord precursor cells at each embryonic stage. (B) The role of *Ci-Bra* in the transcriptional network of *Ciona* embryos. Upstream positive regulators of *Ci-Bra* include maternal β-catenin and *P53*, and zygotic *FoxD*, *FoxA*, *FGF9/16/20*, and *ZicL* (s). *Ci-Bra* directly activates *Ci-tropm* and other notochord-specific genes to induce notochord differentiation (u). Cell polarity-related genes such as *Ci-pk* and *Ci-Fz1/2/7* are involved in convergent extension movements during gastrulation (t). These signaling components are induced directly or indirectly by *Ci-Bra* (v). (Modified from Hotta *et al.* (2008) *Evol Dev*, **10**, 37–51.) (C) Notochord transcription factors, their relationship with *Ci-Bra* (solid arrows) and their putative functions in notochord development (dotted arrows). Notochord cells are depicted in red while the notochordal sheath is shown in purple; white circles represent intercellular lumens. (Adapted from Jose-Edwards *et al.* (2011) *Dev Dyn*, **240**, 1793–1805.)

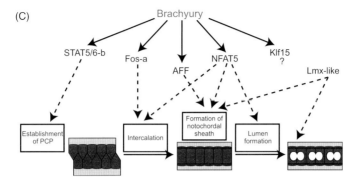

FIGURE 9.3 (*Continued*)

β-catenin mRNA and/or protein that are maternally provided within the egg cytoplasm. After fertilization, cytoplasmic β-catenin protein translocates into the nucleus of vegetal blastomeres, in particular, to those giving rise to endodermal cells, where the protein triggers the expression of several genes, including *FoxA, FoxD*, and *FGF9/16/20* (Fig. 9.3B). In *Ciona* embryos, *FoxD* is expressed in A5.1, A5.2, and B5.1 of the 16-cell-stage embryo (Fig. 9.1b), and in A6.1, A6.3, and B6.1 of the 32-cell-stage embryo (Fig. 9.1c). The 5′ upstream region of *FoxD* contains Tcf-binding motifs, and deletion of these motifs diminishes reporter gene expression in the cells, suggesting that *FoxD* is a direct target of nuclear β-catenin. A Zic family transcription factor gene, *Zic-like (ZicL)* is downstream of *FoxD*, and its zygotic expression commences in A6.2, A6.4, B6.2, and B6.4 of the 32-cell-stage embryo (Fig. 9.1c). ZicL directly binds to its recognition motif located at the 5′ upstream region of *Ci-Bra*, and thus activates the gene in A7.3 and A7.7 at the 64-cell stage (Fig. 9.1d). *FoxA* is also involved in *Ci-Bra* expression via the activation of *ZicL*. A chip–chip analysis confirmed that the *Ciona* notochord GRN consists of interactions between *FoxD/FoxA, ZicL*, and *Bra* (Fig. 9.3B). Therefore, the regulatory cascade leading to *Ci-Bra* expression is β-catenin → *FoxD/FoxA* → *ZicL* → *Bra* (Fig. 9.3B). In addition, the *Ciona* homolog of p53 is maternally provided and activates *ZicL* by binding the p53-recognition site in the upstream region of *ZicL*, thus playing a role in *Bra* activation and notochord differentiation (Fig. 9.3B).

The other, noncell autonomous, signaling cascade that induces *Bra* expression involves FGF signaling between *FGF9/16/20*-expressing endoderm cells and presumptive notochord cells (Fig. 9.2a). A6.2 and A6.4 divide into A7.3/A7.4 and A7.7/A7.8, respectively. A7.3 and A7.7 are located in the middle part of the vegetal hemisphere and become notochord, while A7.4 and A7.8 are located at the anterior side and become neural cells. FGF9/16/20 activates the mitogen activated protein kinase (MAPK)/extracellular signal-regulated protein kinase (ERK) kinase (MEK)–ERK pathway at the 64-cell stage in notochord precursor cells (A7.3/A7.7) but not in nerve-cord precursor cells (A7.4/A7.8). In A7.4/7.8 cells, Ephrin-Eph signals emanating from adjacent a-line cells block the FGF9/16/20-dependent MEK–ERK pathway, and thereby suppress the expression of *Bra* expression in A7.4/A7.8 cells (Fig. 9.2a). *Ciona* embryos in which *FGF9/16/20* function has been depleted eventually to recover the ability to express *Bra*. This and other experimental results suggest that *FGF9/16/20* acts alone during the initial notochord induction step, but acts together with *FGF8/17/18* in the subsequent notochord maintenance step.

As was discussed in several chapters, the molecular mechanisms of embryonic cell specification are not always identical between *Ciona* and *Halocynthia*. In *Ciona*, a microarray analysis revealed that the zygotic expression of *FoxD*, *FoxA*, *ZicL*, and *Bra* occurs even when embryonic cells are fully dissociated from the first cleavage, suggesting a possibility that *Ci-Bra* is able to initiate zygotic expression in the absence of cell–cell interactions up to the 64-cell stage. On the other hand, the initiation of *Bra* transcription in *Halocynthia* embryos is dependent on a cell contact-mediated FGF9/16/20 signaling pathway. In *Halocynthia roretzi*, isolating A6.2 or A6.4 from the 32-cell stage embryo leads to a failure of *Hr-Bra* expression at the 64-cell stage. It has been shown that the FGF–MEK–ERK signaling pathway is essential to *Hr-Bra* expression and that both the Ets- and Zic-binding sites are critical for the initiation of *Hr-Bra* expression.

9.2.3 B-Line Signaling Cascade Upstream of *Bra*

The upstream signaling cascade leading to *Bra* expression in the B-line is distinct from that in the A-line and requires a relay mechanism involving two signaling pathways: first Nodal and then Notch/Delta. As discussed in Chapters 6 and 7, lateral Nodal signaling from adjacent b-line animal cells is required for *Delta-like* expression in A7.6 cells. The Delta-like signaling activates the Notch/Delta pathway in adjacent B-line notochord cells, which in turn activates *Bra* in B8.6 (Fig. 9.2b). In addition, the regulatory repressor Snail acts to form a boundary between B-line muscle (B7.4/B7.8) and notochord/mesenchyme (B7.3/B7.7).

In vertebrate embryos, Sonic hedgehog (Shh) plays a pivotal role in the induction of floor plate by the notochord. The *Ciona intestinalis* genome contains two hedgehog genes (*Ci-hh1* and *Ci-hh2*), which have arisen from a lineage-specific duplication. Neither of the *Ciona hhs* are expressed in the notochord, although *Ci-hh2* is expressed in ventral nerve-cord cells, which are considered to be homologous to the vertebrate floor plate.

9.3 GENES INVOLVED IN NOTOCHORD FORMATION

Approximately 450 genes that are identified through characterization of *Ci-Bra* downstream genes and analyses of mutations leading to abnormal notochord morphogenesis are involved in notochord development. The related experiments are discussed below.

9.3.1 Notochord Genes Downstream of *Ci-Bra*

Subtractive hybridization of mRNA from embryos overexpressing *Ci-Bra* and mRNA from control embryos has identified ~450 candidates as downstream *Ci-Bra* targets expressed in notochord cells (Fig. 9.3). Further characterization of the target genes with respect to genomic and mRNA sequences, and expression patterns as observed by WISH studies reveals that the 450 genes embody a variety of different functions. All of the target gene information is publicly available at CINOBI (http://chordate.bpni .bio.keio.ac.jp/notochord/cibra) and also at http://dev.biologists.org/content/suppl/2010/04 /21/137.10.1613.DC1/046789FigS10.pdf.

WISH analysis demonstrated that 73 genes exhibited the following tissue- or organ-specific expression patterns (Fig. 9.3A): notochord-specific (17 genes), epidermis-specific (17 genes), nervous system-specific (12 genes), endoderm-specific (7 genes), mesenchyme-specific (11 genes), and muscle-specific (9 genes). The remaining genes were expressed in multiple tissues. Genes with notochord-specific expression are likely to be direct targets of *Ci-Bra*. For example, *tensin* (Fig. 9.3c), *zipper/nonmuscle myosin heavy chain II* (Fig. 9.3d), *perlecan* (Fig. 9.3f), and *Collagen alpha type XVIII* (Fig. 9.3g) are notochord-predominant genes and function during notochord formation. In addition, a recent study shows that five transcription factors, S*TAT5/6-b, Fos-a, NFAT5, AFF*, and *Klf15*, are either direct or indirect targets of *Ci-Bra*, with possible roles in various steps of the notochord formation (Fig. 9.3C). In vertebrates, the notochord plays a pivotal role as a source of signals during the development of neighboring tissues and organs. Many of the identified *Ci-Bra* target genes demonstrate non-notochord expression profiles. These genes are likely to represent indirect targets of *Ci-Bra*-induced signaling from the notochord to the surrounding tissues.

9.3.2   Genes Involved in Notochord Cell Morphogenesis

Eleven Wnt/planar cell polarity (PCP genes are found among the downstream *Ci-Bra* genes, which include *Wnt9/14/15*, *Wnt-a* (*Ciona* lineage-specific *Wnt* homolog), *Prickle* (*pk*), *zipper* (Fig. 9.3d), *Dsh, Cdc42, misshapen, STAT, Frizzled1/2/7, Nemo-like kinase* (*NLK*), and *grainy head*. Interestingly, these genes encode a variety of functional components, for example, a receptor (*Fz1/2/7*), ligands (*Wnt9/14/15, Wnt-a*), cytoplasmic modulators (*pk, Cdc42, Dsh*), and an effector (*zipper*). Recently, a link between *Brachyury* and Wnt/PCP signaling has been implicated in vertebrates. *Xenopus Xbra* functions as a switch between cell migration and convergent extension in the gastrula, and *no tail* (a zebrafish *Brachyury* ortholog) co-operates with noncanonical Wnt signaling to regulate posterior body morphogenesis, suggesting that *Brachyury* might regulate the transcription of Wnt/PCP components (Fig. 9.3t).

Ci-SELP (P-selectin) and *Ci-tensin* are also downstream of *Ci-Bra*. P-selectin is a cell adhesion molecule that mediates the adhesion of neutrophils and monocytes to activated platelets and endothelial cells. Tensin is a cytoplasmic phosphoprotein localized to integrin-mediated focal adhesions. It binds to actin filaments and contains a phosphotyrosine-binding (PTB) domain that interacts with the cytoplasmic tails of β-integrins. Tensin stabilizes integrin adhesive contacts in flies. Therefore, the two genes are likely to be involved in the adhesion and migration of notochord cells.

9.3.3   Genes Involved in the Structural Formation of the Notochord

As development progresses, the notochord is surrounded by the notochord sheath, an extracellular matrix structure. Four collagen family genes, *Ci-collagen XI* (Fig. 9.3h), *Ci-collagen XVIII* (Fig. 9.3g), *Ci-collagen II*, and *Ci-collagen XXII*, are downstream of *Ci-Bra*. Of these, *Ci-collagen XVIII*, *Ci-collagen II*, and *Ci-collagen XI* are expressed in the notochord. Collagen XVIII is a basement membrane-associated, heparan sulfate proteoglycan. Two additional proteoglycans that are involved in notochord development, leprecan and perlecan (Fig. 9.3f), are also downstream of *Ci-Bra*. The leucine proline-enriched proteoglycan leprecan is a chondroitin sulfate proteoglycan. Perlecan is a large basement membrane-associated heparan sulfate proteoglycan. The combination of collagen (type II) and proteoglycans in the notochord and cartilage is a conserved feature of higher vertebrates (see Chapter 17), and are also found in the lamprey notochord sheath. These extracellular matrix components are involved in the structural formation of the notochord or the notochord sheath (Fig. 9.3u), thus justifying their conservation as notochord constituents among chordates. *Ci-myosin I*β, *Ci-IQGAP*, *Ci-myosin light chain alkali*, *Ci-NCAM* (Fig. 9.3j), and *Ci-ciboulot* (*thymosin beta*) (Fig. 9.3i) are also likely to be involved in notochord structural formation.

9.3.4   Genes Involved in Notochord Induction of CNS Differentiation

In chordates, especially in vertebrates, the notochord has been implicated in inducing the formation of CNS structures, including the floor plate, the paraxial mesoderm (e.g., somites), and derivatives of the endoderm. Genes downstream of *Ci-Bra*, such as *Ci-netrin* and *Ci-fibrinogen-like* (*Ci-fibrn*) serve as inductive signals from the notochord, which lead to the expression of genes required for the differentiation of other tissues. For example, the expression of *Ci-ciboulot, Ci-GATAa, Ci-NCAM,* and *Ci-Snail* is influenced by *Ci-netrin* and *Ci-fibrinogen-like*. Genes for sequence-specific DNA-binding proteins including *Ci-Dlx-c, Ci-FoxC, Ci-Snail, Ci-AP4,* and *Ci-STAT5/6-b* are also characterized as downstream of *Ci-Bra* (Fig. 9.3v). These genes may function in the formation of other tissues and organs, as they are not expressed in the notochord.

Ci-fibrn is of special interest with respect to notochord–CNS interactions (Fig. 9.4). *Ci-fibrn* encodes a fibrinogen-like protein. The gene is expressed specifically in notochord cells from the neurula stage (Fig. 9.4a); however, its protein product is not confined to those cells but is distributed underneath the nervous system as fibril-like protrusions in developing tailbud embryos (Fig. 9.4b, d, e). At the same stage, cells of the CNS express Notch (Fig. 9.4e), which is essential for the correct distribution of Ci-fibrn protein. A disturbance in Ci-fibrn distribution produces a failure in dorsal positioning of neuronal cells and an abnormal axon extension path. Therefore, the physical and functional interactions between the notochord-based fibrinogen-like protein and the neural tube-based Notch are essential for the dorsal patterning of the nervous system, and—more broadly—for the establishment of the chordate body plan (Fig. 9.4f).

9.3.5   Genes Identified by Mutant Analyses and Morphogenetic Mechanisms

As in vertebrate embryos, the PCP pathway is essential for ascidian notochord formation. Disheveled (Dsh) is required for proper notochord cell intercalation. Prickle (pk) also functions in notochord morphogenesis, because *aimless*, a null mutation of pk, severely disrupts notochord morphogenesis. In normal embryos, the Dsh protein is membrane-associated except at the notochord-muscle boundary, where it is excluded. Thus, Dsh localization displays mediolateral polarization. In *aimless* embryos Dsh membrane localization is disrupted, and mediolateral polarization of notochord cells is lost.

Another recessive short-tailed *Ciona savignyi* mutation *chongmague* has a novel defect in the formation of a morphological boundary around the developing notochord. In *chongmague* embryos, notochord cells initiate intercalation normally, but then fail to maintain their polarized cell morphology and migrate inappropriately to become dispersed in the larval tail. Positional cloning identifies *chongmague* as a mutation in the *C. savignyi* ortholog of the vertebrate alpha 3/4/5 family of laminins. *Cs-lam3/4/5* is highly expressed in the developing notochord, and its protein is specifically localized to the outer border of the notochord. Notochord convergence and extension, reduced but not absent in both mutants, are essentially abolished in the *aim/aim : chm/chm* double mutant, indicating

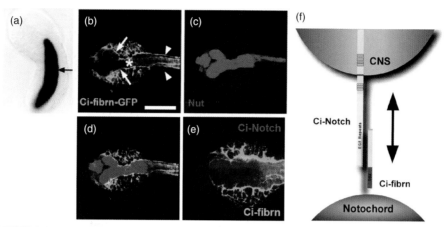

FIGURE 9.4   Notochord-specific gene *Ci-fibrn* encodes a fibrinogen-like protein that interacts with Notch in the CNS to regulate neuronal cell patterning. (a) Specific expression of the *Ci-fibrn* gene in notochord cells (arrow) as revealed by whole-mount *in situ* hybridization. (b–d) Extracellular distribution of Ci-fibrn protein. Expression of EGFP and RFP reporters in an embryo injected with (b) Ci-Bra(notochord)-promoter:fibrn-EGFP (green), and (c) Ci-Nut(CNS)-promoter:RFP (magenta). (d) Merged photomicrograph of (b) and (c). Scale bar, 50 μm. (e) Expression of Ci-fibrn (green) and Ci-Notch (magenta) interactions. (f) Diagram illustrating the contribution of Notch signaling to Ci-fibrn function. (Adapted from Yamada *et al.* (2009) *Dev Biol*, **328**, 1–12.)

that laminin-mediated boundary formation and PCP-dependent mediolateral intercalation are each able to drive a remarkable degree of tail morphogenesis in the absence of the other. These mechanisms initially act in parallel, but PCP signaling also has an important later role in maintaining the perinotochordal/intranotochordal polarity of Cs-lam3/4/5 localization.

Finally, *Ciona* provides an appropriate experimental system to explore the cellular mechanisms involved in cell elongation, a fundamental process that allows cells and tissues to adopt new shapes and functions (Fig. 9.5). During *Ciona* notochord tubulogenesis, a dramatic elongation of individual cells takes place that lengthens the notochord and, consequently, the entire embryo. A recent study found a novel dynamic actin- and nonmuscle myosin II-containing constriction midway along the anteroposterior aspect of each notochord cell during this process (Fig. 9.5c). Both actin polymerization and myosin II activity are required for the constriction and for cell elongation. Following elongation, the notochord cells undergo a mesenchymal–epithelial transition and form two apical domains at opposite ends (Fig. 9.5d). Extracellular lumens then form at the apical surfaces. Cortical actin and ezrin–radixin–moesin (ERM) proteins are essential for lumen formation, and a polarized network of microtubules, which contributes to lumen development, forms in an actin-dependent manner at the apical cortex. Later in notochord tubulogenesis, when the notochord cells initiate bidirectional crawling along the notochord sheath, the microtubule network rotates 90° and becomes organized as parallel bundles extending toward the leading edges of tractive lamellipodia (Fig. 9.5e). This process is required for the correct organization of actin-based protrusions and subsequent lumen coalescence. As a model system, *Ciona* has played a role in describing the contributions of the actomyosin and microtubule networks to notochord tubulogenesis, and has revealed the extensive crosstalk and regulation between these two cytoskeletal components.

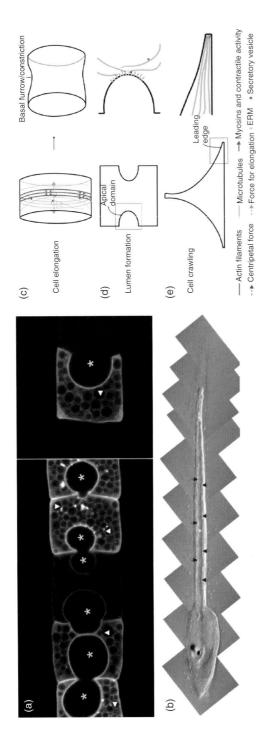

FIGURE 9.5 Morphogenesis events during notochord formation in *Ciona* embryos. (a) Extracellular pockets in the notochord column. A membrane-targeted green fluorescent protein (GFP; white in this image) was introduced by electroporation and is expressed in notochord cells in a mosaic fashion. The asterisk indicates extracellular pockets. Arrowheads indicate intracellular yolk granules. Anterior to the left. (b) A *C. savignyi* swimming tadpole imaged by Normarski microscopy. At this stage, the notochord is a tubular lumen at the center (white asterisk) encircled by endothelial-like notochord cells in the periphery (arrows). The notochord plays an important role in locomotion by serving as a hydrostatic skeleton in the larval tail. Lateral view. (Adapted from Jiang and Smith (2007) *Dev Dyn*, **236**, 1748–1757.) (c–e) Models for cytoskeleton organization at three stages of notochord tubulogenesis in *Ciona*. (c) Actin and myosin II form an actomyosin network at the equatorial region of the basal cortex. The contractile activity of this network generates compression along the cell circumference and causes a furrow to form at the equator. The basal contractions presumably produce a centripetal force, which is converted to a pushing force exerted on both the anterior and the posterior surfaces of the notochord cell, driving the cell to elongate. (d) Apical cortical actin, together with microtubules, transport luminal vesicles that contribute to lumen formation. Ezrin/radixin/moesin (ERM) proteins are essential for luminal membrane specification and lumen formation. (e) Microtubules form bundles that extend to the leading edges of lamellipodia as the notochord cells crawl. Microtubule extension at the cell edge might provide a mechanical/positional cue for the orientation of protrusions. (Adapted from Dong *et al.* (2011) *Development*, **138**, 1631–1641.)

In summary, *Ciona* provides a system in which the cellular and molecular mechanisms involved in notochord formation can be extensively elucidated. Further studies will uncover more mechanistic details leading to increasingly accurate descriptions of the underlying GRN.

SELECTED REFERENCES

CORBO, J.C., LEVINE, M., ZELLER, R.W. (1997) Characterization of a notochord-specific enhancer from the *Brachyury* promoter region of the ascidian, *Ciona intestinalis*. *Development*, **124**, 589–602.

DONG, B., DENG, W., JIANG, D. (2011) Distinct cytoskeleton populations and extensive crosstalk control *Ciona* notochord tubulogenesis. *Development*, **138**, 1631–1641.

HOTTA, K., TAKAHASHI, H., SATOH, N., GOJOBORI, T. (2008) *Brachyury*-downstream gene sets in a chordate, *Ciona intestinalis*: integrating notochord specification, morphogenesis and chordate evolution. *Evol Dev*, **10**, 37–51.

IMAI, K., TAKADA, N., SATOH, N., SATOU, Y. (2000) β-catenin mediates the specification of endoderm cells in ascidian embryos. *Development*, **127**, 3009–3020.

IMAI, K.S., SATOH, N., SATOU, Y. (2002a) An essential role of a *FoxD* gene in notochord induction in *Ciona* embryos. *Development*, **129**, 3441–3453.

IMAI, K.S., SATOH, N., SATOU, Y. (2002b) Early embryonic expression of *FGF4/6/9* gene and its role in the induction of mesenchyme and notochord in *Ciona savignyi* embryos. *Development*, **129**, 1729–1738.

IMAI, K.S., SATOU, Y., SATOH, N. (2002c) Multiple functions of a Zic-like gene in the differentiation of notochord, central nervous system and muscle in *Ciona savignyi* embryos. *Development*, **129**, 2723–2732.

IMAI, K.S., LEVINE, M., SATOH, N., SATOU, Y. (2006) Regulatory blueprint for a chordate embryo. *Science*, **312**, 1183–1187.

JIANG, D., MUNRO, E.M., SMITH, W.C. (2005) Ascidian prickle regulates both mediolateral and anterior–posterior cell polarity of notochord cells. *Curr Biol*, **15**, 79–85.

JIANG, D., SMITH, W.C. (2007) Ascidian notochord morphogenesis. *Dev Dyn*, **236**, 1748–1757.

JOSE-EDWARDS, D.S., KERNER, P., KUGLER, J.E., DENG, W. *et al.* (2011) The Identification of transcription factors expressed in the notochord of Ciona intestinalis adds new potential players to the Brachyury gene regulatory network. *Dev Dyn*, **240**, 1793–1805.

KIM, G.J., KUMANO, G., NISHIDA, H. (2007) Cell fate polarization in ascidian mesenchyme/muscle precursors by directed FGF signaling and role for an additional ectodermal FGF antagonizing signal in notochord/nerve cord precursors. *Development*, **134**, 1509–1518.

KUBO, A., SUZUKI, N., YUAN, X., NAKAI K. *et al.* (2010) Genomic cis-regulatory networks in the early *Ciona intestinalis* embryo. *Development*, **137**, 1613–1623.

KUMANO, G., YAMAGUCHI, S., NISHIDA, H. (2006) Overlapping expression of *FoxA* and *Zic* confers responsiveness to FGF signaling to specify notochord in ascidian embryos. *Dev Biol*, **300**, 770–784.

MATSUMOTO, J., KUMANO, G., HIROKI NISHIDA, H. (2007) Direct activation by Ets and Zic is required for initial expression of the *Brachyury* gene in the ascidian notochord. *Dev Biol*, **306**, 870–882.

MUNRO, E.M., ODELL, G.M. (2002) Polarized basolateral cell motility underlies invagination and convergent extension of the ascidian notochord. *Development*, **129**, 13–24.

NAKATANI, Y., MOODY, R., SMITH, W.C.. (1999) Mutations affecting tail and notochord development in the ascidian *Ciona savignyi*. *Development*, **126**, 3293–3301.

PICCO, V., HUDSON, C., YASUO, H. (2007) Ephrin-Eph signalling drives the asymmetric division of notochord/neural precursors in *Ciona* embryos. *Development*, **134**, 1491–1497.

SATOH, N., TAGAWA, K., TAKAHASHI, H. (2012) How was the notochord born? *Evol Dev*, **14**, 56–75.

SATOU, Y., SATOH, N., IMAI, K.S. (2009) Gene regulatory networks in the early ascidian embryo. *Biochim Biophys Acta*, **1789**, 268–273.

TAKAHASHI, H., HOTTA, K., ERIVES, A., DI GREGORIO A. *et al.* (1999) *Brachyury* downstream notochord differentiation in the ascidian embryo. *Genes Dev*, **13**, 1519–1523.

TAKATORI, N., SATOU, Y., SATOH, N. (2002) Expression of hedgehog genes in *Ciona* intestinalis embryos. *Mech Dev*, **116**, 235–238.

VEEMAN, M.T., NAKATANI, Y., HENDRICKSON, C., ERICSON, V. *et al.* (2008) Chongmague reveals an essential role for laminin-mediated boundary formation in chordate convergence and extension movements. *Development*, **135**, 33–41.

YAGI, K., SATOU, Y., SATOH, N. (2004) A zinc finger transcription factor, ZicL, is a direct activator of Brachyury in the notochord specification of *Ciona intestinalis*. *Development*, **131**, 1279–1288.

YAMADA, S., HOTTA, K., YAMAMOTO, T.S., UENO, N. *et al.* (2009) Interaction of notochord-derived fibrinogen-like protein with Notch regulates the pattering of the central nervous system of *Ciona intestinalis* embryos. *Dev Biol*, **328**, 1–12.

YASUO, H., SATOH, N.. (1993) Function of vertebrate T gene. *Nature*, **364**, 582–583.

YASUO, H., HUDSON, C. (2007) FGF8/17/18 functions together with FGF9/16/20 during formation of the notochord in *Ciona* embryos. *Dev Biol*, **302**, 92–103.

CHAPTER 10

THE LARVAL AND ADULT NERVOUS SYSTEMS

Ascidian larvae display complex motile behaviors in response to light and gravity, reactions that are governed by well-organized CNS and PNS. Nevertheless, the larval nervous system is quite simple compared to that of vertebrates. The larval CNS of *Ciona intestinalis* consists of ~400 cells, approximately 180 being neurons, while the remainder are thought to be nonneuronal ependymal or glial cells. The PNS contains ~20 epidermal mechanosensory neurons. Taking advantage of the structural simplicity and genomic information available in this model organism, the predominant mechanisms underlying larval CNS and PNS development have been described at the single cell level. This chapter also discusses the adult ascidian nervous system, which forms the so-called neural complex comprising of the cerebral ganglion and the neural gland.

10.1 LARVAL CNS

10.1.1 Tripartite or Bipartite Organization of the CNS

Morphologically the ascidian larval CNS is divided into five major parts along the A–P axis: the sensory vesicle, posterior brain, neck, visceral (motor) ganglion, and tail nerve cord (Fig. 10.1a and Fig. 10.5a). As the vertebrate CNS uses a tripartite organization consisting of the fore, mid, hindbrain, and spinal cord, the regional homology between ascidians and vertebrates has been investigated. The expression pattern of *Otx*, *Pax2/5/8*, *FGF8/17/18*, *En*, and *Hox1* in the ascidian larval CNS suggests that it may be mapped using a tripartite organization (Fig. 10.3f). Specifically, the sensory vesicle is homologous to the vertebrate forebrain, the neck likely homologous to the vertebrate midbrain–hindbrain boundary (MHB), and the visceral ganglion and tail nerve cord are homologous to the vertebrate hindbrain and spinal cord.

Investigations into the expression of *Otx*, *Pax2/5/8*, *FGF8/17/18*, *En*, and *Hox1* in the developing *Ciona* CNS revealed that their expression profiles are quite dynamic at the early and mid tailbud stages. For example, at the early tailbud stage there is no gap between the *Otx*- and *Hox1*-expressing regions and the *Pax2/5/8*-expressing region overlaps with the anterior portion of the *Hox1*-expressing region (Fig. 10.3f). As development progresses, the *Pax2/5/8*-expressing region is then separated from the *Hox1*-expressing region. These observations suggest that the tripartite organization of *Ciona* larval CNS is derived from a "bipartite" organization of CNS, which is characterized by the *Otx*- and *Hox1*-expressing regions. The CNS of larvaceans and cephalochordates display a

Developmental Genomics of Ascidians, First Edition. Noriyuki Satoh.
© 2014 John Wiley & Sons, Inc. Published 2014 by John Wiley & Sons, Inc.

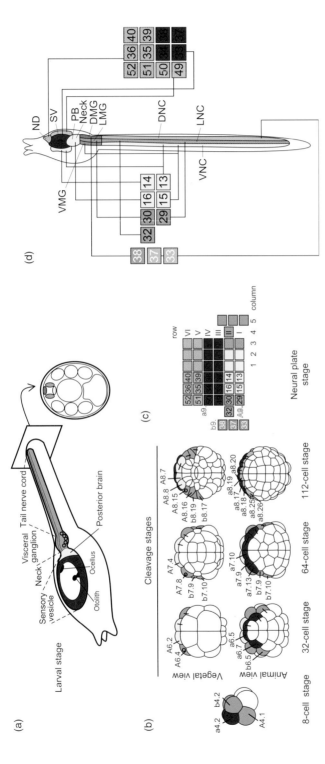

FIGURE 10.1 Cell lineages of the ascidian larval CNS. (a) A diagram showing the structure of the CNS, which consists of the sensory vesicle (red), posterior brain (yellow), neck (yellow), visceral (motor) ganglion (green and brown), and tail nerve cord (green and brown). Small circles indicate motoneurons in the visceral ganglion. (b) Lineages: the a-line is colored red (anterior sensory vesicle precursors) or pink (anterior epidermis and pharynx/neurohypophysis precursors); the b-line is green; and the A-line is yellow at the 8-cell stage and light yellow (medial cells) or tan (lateral cells) from the 32-cell stage. Bars connecting two blastomeres on the right side of the drawings indicate a sister cell relationship. (c) The neural plate of the neural plate stage embryo. a-line cells form the anterior of the neural plate, while A- and b-line cells form the posterior and lateral portions of the neural plate. The plate consists of six (I–VI) rows and five (1–5) columns. The rows and columns of the neural plate are indicated by the Roman and Arabic numerals, respectively. Adapted from Hudson and Yasuo (2005), *Development*, **113**, 2855–2864.) (d) Relationship between neural plate cells and their final destination. Only the names of the left or right blastomere are shown. DNC, dorsal nerve cord; DMG, dorsal motor ganglion; LNC, lateral nerve cord; LMG, lateral motor ganglion; ND, neurohypophysial duct; PB, posterior brain; SV, sensory vesicle; VNC, ventral nerve cord; VMG, ventral motor ganglion. (Adapted from Sasakura *et al.* (2012) *Dev Growth Differ*, **54**, 420–437.)

similar expression pattern of *Otx* and *Hox1*, suggesting that the two groups also use a bipartite organization for their CNS. Therefore, the ascidian CNS is likely to represent an intermediate stage between the bipartite legacy of the common chordate ancestor and the tripartite organization of the vertebrates.

The vertebrate MHB organizer secretes Wnt1 and FGF8 as its organizing activities. Although the *Ciona* genome does not contain *Wnt1*, an ortholog of *FGF8* is expressed in a localized fashion during *Ciona* CNS development at the late gastrula stage and at the mid-tailbud stage (Fig. 10.3f). It should be emphasized that *FGF8* plays a pivotal and conserved role in the MHB organizer in both ascidians and vertebrates.

10.1.2   Lineage and Fate Determination of CNS cells

The CNS originates from a-, b-, and A-line blastomeres (Fig. 10.1b). The simple but beautiful organization of cells in the neural plate appears as a grid comprised of six rows along the A–P axis and five bilateral pairs of columns (Fig. 10.1c). As shown in Fig. 10.1b, the lineage of these neural plate cells is completely documented. The *Ciona* neural plate consists of 44 neural cells in total, 24 of which are derived from the a-line, 14 from the A-line, and the remaining six from the b-line (Fig. 10.1b, c).

The sensory vesicle is formed by a-line cells. The three columns of a9.37/38, a9.33/34, and a9.49/50 give rise to the sensory vesicle, while a9.39/40, a9.35/36, and a9.51/52 form the neurophysical duct, which generates a part of the adult CNS after metamorphosis (Fig. 10.1d). The cells of the posterior brain, neck, and the lateral and ventral rows of the visceral ganglion, and tail nerve cord are derived from A-line cells. Specifically, these are A9.13/14, A9.15/16, A9.29/30, and A9.32 of rows I and II of the neural plate (Fig. 10.1c, d). A9.13/14 produces a ventral row of the nerve cord, and A9.15/16 yields the posterior brain and neck region (Fig. 10.1d). A9.29 and A9.32 form lateral rows of nerve cord, and A9.30 makes the lateral motor ganglion (Fig. 10.1d). The cells comprising the dorsal row of the viscera ganglion and tail nerve cord originate from b-line cells. The lateral b9.33 and b9.37/38 form the dorsal row of nerve cord (Fig. 10.1d).

10.1.3   Overall Specification Mechanisms

10.1.3.1   Forming the Neural Plate   The specification mechanisms differ among a-, b-, and A-line neural cells. The a-line-derived sensory vesicle expresses *Otx*. When *Otx* expression is used as a marker, FGF9/16/20, which is expressed in vegetal cells or mesendoderm precursor at the 16- and 32-cell stages, acts as an endogenous neural inducer (Fig. 10.2a). Two maternally expressed transcription factors, Ets1/2 and GATA-a, work via FGF/MEK/ERK signaling pathway to activate *Otx* transcription. On the other hand, Otx recognizes *cis*-regulatory elements that contain duplicated Otx binding motifs to activate genes responsible for the anterior neuroectoderm. b-line neural tissue is induced by a similar mechanism. However, A-line neural cells are specified quite differently and requires suppression of FGF signaling in order for them to adopt neural fate.

At the 32-cell stage, FGF9/16/20 activates *Nodal* expression in b6.5 via MEK/ERK signaling (Fig. 10.2a, b). Nodal signaling from b6.5 is required for lateral neural plate cell fates (columns 3 and 4 of the adjacent A6.4-line and 5 of the b-line) (Fig. 10.2c, d) and restricts the medial fates (columns 1 and 2) (Fig. 10.2d). Next, the Delta2/Notch pathway acts to specify column 4 fates and a second, later Delta2/Notch signal specifies column 2

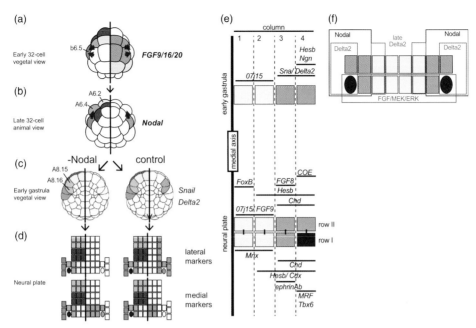

FIGURE 10.2 Signaling pathways involved in the early phase of ascidian larval CNS cell specification leading to neural plate cell patterning. (a–d) The role of FGF9/16/20 and Nodal. Embryonic stage is indicated to the left of the drawings. On the left half of each drawing, the neural lineages are indicated using the same color code as in Fig. 10.1. On the right half of each drawing, blastomeres expressing the markers indicated to the right are shown in grey. From the early gastrula stage, drawings are shown in two columns; those on the right indicate control embryos and those on the left represent embryos in which Nodal signaling has been inhibited. Thick blue arrows represent signaling between blastomeres. (Adapted from Hudson and Yasuo (2005), *Development*, **132**, 1199–1210.) (e, f) Schematic representation of the gene expression profiles in A-line neural lineages at the early gastrula and neural plate stages. The right-hand-side A-line neural blastomeres of the bilaterally symmetrical embryo are represented; the position of the medial axis is indicated on the left. Bars above and below the schematics indicate gene expression for the markers (row II above, row I below). Position of the columns and rows are indicated (top and right). Sna, *Snail*; FGF8, *FGF8/17/18*; Chd, *Chordin*; and FGF9, *FGF9/16/20*. (f) The overlapping requirements of Nodal, Delta2 and FGF/MEK/ERK signalling pathways in the A-line neural plate. Nodal first separates the A-line neural plate cells into medial (column 1 and 2) and lateral (column 3 and 4) domains. Delta2/Notch then first specifies column 4 fates at the 76-cell stage, then at the early gastrula stage, a second phase of Delta2 signaling distinguishes columns 1 and 2. At the mid-gastrula stage, ERK is activated throughout rows I and III. Activation in row I is required for the distinction between rows I and II. (Adapted from Hudson *et al.* (2007) *Development*, **134**, 3527–3537.)

fates (Fig. 10.2e, f). Finally, along the anterior–posterior axis, FGF/MEK/ERK promotes row I fates and restricts row II fates in the A-line neural plate cells (Fig. 10.2e, f). Thus, at least three distinct signaling pathways—FGF/MEK/ERK, Nodal, and Delta2/Notch pathways—are involved in the patterning of the A-line neural plate cells.

10.1.3.2 The Compartmentalization of the Ciona Larval CNS
The compartmentalization of the *Ciona* larval CNS is accomplished by a regulatory network of transcription factors and signaling pathway molecules (Fig. 10.3). First, a localized

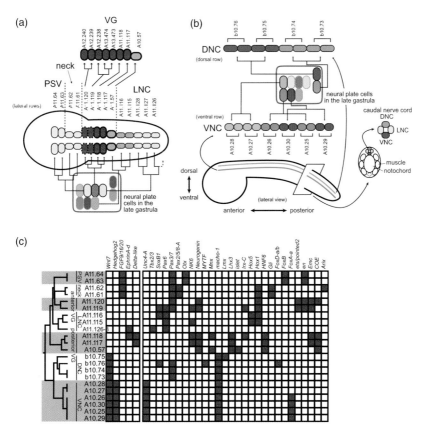

FIGURE 10.3   Development of the A-line central nervous system in the *Ciona* embryo. (a, b) A schematic representation of the CNS cells and their lineages at the tailbud stage from the (a) dorsal and (b) lateral views. Arrows indicate cell lineages. Cells of the same color are derived from a single cell at the late gastrula stage (enclosed by pink lines). Cells enclosed by thick black lines are postmitotic cells destined to become motoneurons. At the early tailbud stage, three pairs of presumptive motoneurons are postmitotic. By the mid-tailbud stage, two pairs of postmitotic presumptive motoneurons are differentiated from the remaining two pairs of visceral (motor) ganglion (VG) cells, as shown in the upper part of (b). (c) Expression profiles of the regulatory genes. A- and b-line neural cells in the posterior sensory vesicle (PSV), neck, VG, and nerve cord of the early tailbud stage. (left side) A hierarchical clustering of the neural cells based on the expression profiles of transcription factor genes. (d, e) Gene regulatory network information of the *Ciona* CNS. (d) Summary of the networks. (e) Diagrams of gene circuits in A11.64 (PSV), A11.62 (neck), A11.118 (VG), and A11.116 (caudal nerve cord) at the early tailbud stage. The cell color code is the same as in (a). Transcription factor genes and signaling ligand genes are indicated by rectangles and ovals, respectively. Genes that are expressed in precursors of the cell but are no longer expressed at this stage are enclosed by broken lines. Genes that are not expressed in either the specified cell or its precursors are shown in gray. Arrows indicate transcriptional regulatory interactions. Lines ending in a bar indicate inhibition. (f) Schematic representations of the brain regionalization mechanism by *FGF8/17/18* in (left) the normal embryo and (right) *FGF8/17/18*-morphant embryo (i.e., embryo in which *FGF8/17/18* expression has been downregulated by treatment with morpholino oligomers). Note that *FGF8/17/18* is not expressed in the VG at the tailbud stage but is expressed in the progenitor cells (A9.30) within the neural plate. (g) Schematic diagram illustrating the regulation of *FGF8/17/18* expression in A9.30. (Adapted from Imai *et al.* (2009) *Development*, **136**, 285–293.)

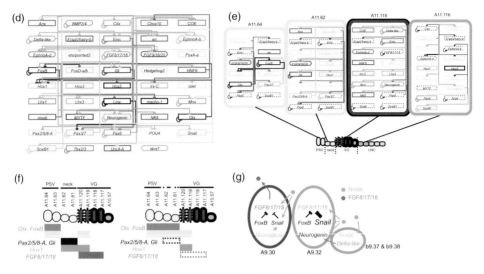

FIGURE 10.3 (*Continued*)

FGF8/17/18 signaling center is likely to be used to delineate the neck and visceral ganglion of the *Ciona* larval brain. Second, FGF8/17/18 signaling in *Ciona* is required to restrict expression of *Otx* and *FoxB* to the posterior brain, as well as being required for expression of *Pax2/5/8-A* in the neck (Fig. 10.3f, g). *Otx* and *FoxB* might inhibit *Hox1* expression in the forebrain via *Cyp26*, a gene encoding an enzyme for RA (retinoic acid) degradation, whereas *Pax2/5/8-A* might coordinate the expression of the regulatory genes required for the differentiation of the motoneurons, such as *Phox2a/Arix*. Finally, localized FGF8/17/18 signaling influences the expression of *Pax2/5/8-A* (Fig. 10.3f, g).

The nerve cord consists of only four rows of cells: one dorsal, one ventral, and two lateral rows (Fig. 10.1a, d). The dorsal row originates, from anterior to posterior, from the following b-line cells: b10.76, b10.75, b10.74, and b10.73 (Fig. 10.1 and Fig. 10.3b). The ventral row originates, from anterior to posterior, from the following A-line cells: A10.28, A10.27, A10.26, A10.30, A10.25, and A10.29 (Fig. 10.3b). The lateral rows originate from the following A-line cells: A11.116, A11.115, A11.126, A11.127, and A11.128 (Fig. 10.3b). Each line of progeny uses a specific combination of regulatory genes (Fig. 10.3c–e). In this respect, the ventral row shares similarity with the vertebrate floor plate (e.g., both express *HNF-3b/forkhead* and *Sonic hedgehog*).

10.1.3.3 Neural Tube Formation
Ascidian neurulation takes place in a similar manner to vertebrates, by rolling up the neural plate at the dorsal side to form a tubular structure that sinks beneath the epidermal layer. Several gene activities are required for proper neural tube function. For example, the movement of the midline epidermal cells to close the neural tube is dependent on the action of actin filaments as mediated via Rho/ROCK (Rho-associated kinase) signaling. The medial plane of the midline epidermis accumulates actin filaments during this movement.

In addition to the mediolateral patterning of the neural plate, Nodal signaling is thought to be responsible for *Ciona* neurulation, as disrupting Nodal function in cleaving embryos leads to neural tube malformation and failure to close. Nodal activates the expression of two transcription factor genes, *ZicL* and *Cdx*, in the neural plate. Microarray analyses have shown that genes encoding extracellular matrix factors and protocadherins are targets of Nodal signaling in the neural tissue. In addition, *Prickle*—a known regulator of planar cell polarity—is also upregulated by Nodal signaling. These data indicate a significant role for Nodal signaling in CNS patterning by directly regulating the morphogenetic movements of the neural plate cells during neurulation.

Changes in the cell cycle are also associated with neural tube formation. At the cleavage stage, the epidermal cell cycle displays very short gap phases or none at all. However, a long G2 phase is inserted into the epidermal cell cycle at the time of neural tube closure. The epidermal cells move toward the midline during the G2 phase, and undergo mitosis immediately after the completion of neural tube closure. Cdc25 is a conserved cell cycle regulator that facilitates both G1/S and G2/M transitions. Regulating the expression of *Cdc25* is crucial for the elongation of the epidermal G2 phase, and shortening the G2 phase by artificially inducing *Cdc25* expression leads to the failure of neural tube closure.

10.1.3.4 *Specification of the Larval Motor Neurons*

The mechanisms that specify five pairs of cholinergic motor neurons in the motor ganglion have been extensively studied. From anterior to posterior the five pairs are aligned as A12.239 (a daughter of A11.120), A13.474 (a daughter of A11.119), A11.118, A11.117, and A10.57 (Fig. 10.3a and Fig. 10.4b). As shown in Fig. 10.3a, the lineage of all five pairs is completely documented. The most anterior pair, the A12.239 neurons, extends axons contralaterally (Fig. 10.4b). The axons of the A11.118 and A10.57 pairs extend their endplates toward the tail muscle, suggesting that the neurons innervate this muscle (Fig. 10.4b). The neuronal pairs A11.117 and A13.474 extend their axons toward the tail, but they do not have endplates that extend toward the muscle (Fig. 10.4b).

Each bilateral pair of neurons expresses different sets of transcription factors. *Dmbx* is expressed only in the A12.239 pair; *Vsx* is initially made in A11.117, and later in A13.474; *Pitx* is produced only in A11.117; *Nk6* is found in A11.118; and *Pitx* and *Islet* are expressed only in A10.57 (Fig. 10.3c). The vertebrate orthologs of these genes play essential roles in motor neuron and interneuron formation in the spinal cord. Disrupting the expression pattern of these regulatory genes changes the character of the neuron pairs in a way that corresponds to the expression of the regulatory genes. The expression patterns of these transcription factors in the motor ganglion are achieved through FGF and Notch signaling. For example, in response to signals from FGF8, A10.60 differentiates into A11.120 and A11.119 (Fig. 10.4c), and in response to Notch signaling, A11.119 differentiates into A12.236, A13.474, and A13.473 (Fig. 10.4d), suggesting that very local cell–cell communication governs the patterning of the motor ganglion at a single cell resolution.

In vertebrates, BMP2/4 signaling and Hedgehog play crucial roles in motoneuron formation. In *Ciona*, however, neither BMP2/4 signaling nor the ventral cell lineages expressing hedgehog play crucial roles in motoneuron formation. In addition, BMP2/4 overexpression induced ectopic motoneurons, the opposite of its vertebrate role. Therefore, it is likely that the specification mechanisms of motoneurons has been modified in the two lineages after their evolutionary split, such that BMP2/4 has adopted a redundant inductive role rather than a repressive role and Nodal that is expressed upstream of BMP2/4 in the dorsal neural tube precursors acts as a motoneuron inducer during normal development.

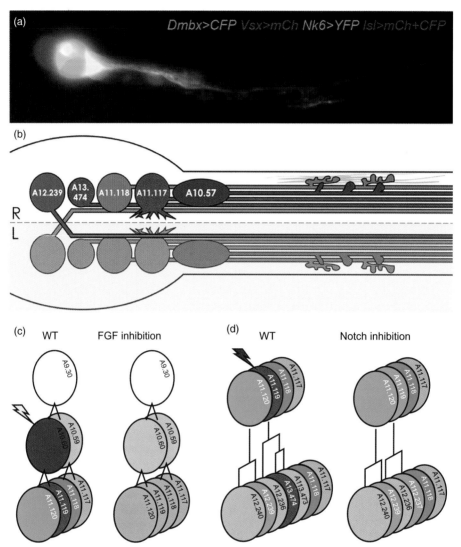

FIGURE 10.4 The *Ciona* visceral ganglion "brainbow." (a) Merged view of an imaged larva electroporated with a combination of the following plasmids: Dmbx>CFP (blue), Vsx>mCherry (red), Nk6>YFP (green), and Islet>mCherry with Islet>CFP (purple). As five pairs of motoneuron express *Dmbx, Vsx, Nk6, Islet*, and *Pitx/Islet* each, all the motoneurons are simultaneously visualized and distinguishable from one another. Yellow indicates overlap of cell bodies, not colocalization. (b) The neurons depicted in (a) and their unique morphological traits, such as contralateral projection of the A12.239 pair, the smaller cell body of A13.474, the frondose endplates of A11.118, the dendrites on A11.117, and the elongated cell body and smaller endplates of A10.57. Pale-green shading indicates the left side of the larva. Pink fibers represent the tail muscle. (c) The mitotic history of the A9.30 lineage, indicating putative FGF signaling events (yellow thunderbolts) distinguishing A10.60 from A10.59, and A11.118 from A11.117 in wild-type embryos (left). Inhibiting FGF signaling (right) transforms the entire lineage to an A11.117-like fate (orange). (d) The mitotic history of the A9.30 lineage. The pink thunderbolt indicates that upon inhibition of Notch A11.119 adopts an A11.120 fate, suggesting that Notch signaling is required for the specification of A11.119. When Notch signaling is inhibited, A11.119 adopts an A11.120 fate, giving rise to an ectopic A12.239-like descendant (light blue). (Adapted from Stolfi and Levine (2011) *Development*, **138**, 995–1004.)

10.1.3.5   Neuronal Cell Types and Neurotransmitter   The classification of neural cell types is essential in order to understand the function of the neural network. Changes in membrane potential affect ion channels and transporters, which then alter intracellular chemical conditions. A genome-wide analysis of the ion channel genes of *C. intestinalis* identified a novel protein with a voltage-sensing transmembrane domain homologous to the S1–S4 segments of voltage-gated channels and a cytoplasmic domain similar to phosphatase and tensin homologues. This protein, named *C. intestinalis* voltage-sensor-containing phosphatase (Ci-VSP), can affect channel-like "gating" currents and directly translates changes in membrane potential into the turnover of phosphoinositides. The phosphoinositide phosphatase activity in Ci-VSP acts within a physiological range of membrane potential. Thus, voltage sensing can lead to functions beyond the conventionally understood ion channel proteins, and may have a broader functional impact than had been previously realized.

Neurotransmitter vesicular transporters and key enzymes involved in neurotransmitter synthetic pathways make good neuronal subtype-specific markers. In addition, the distribution of neurotransmitter receptors has been used as a proxy to identify neurons that could respond to particular neurotransmitters. In *C. intestinalis*, the neurotransmitter properties of most larval neurons have been documented. The larva contains most of the major neuronal types observed in the vertebrate brain: dopaminergic, serotonergic, glutamatergic, cholinergic, GABAergic/glycinergic, and peptidergic neurons. The presence of noradnreregic neurons has also been suggested in the *C. savignyi* larvae. Several of the *Ciona* genes are described below (see Horie *et al.* (2009) for additional descriptions).

Glutamatergic neurons. Glutamate is a major neurotransmitter in the excitatory synapses of both vertebrate and invertebrate nervous systems. In the vertebrate CNS, glutamatergic neurons work in different neural processes including photo-, mechano-, and chemosensations, neural development, motor control, learning, and memory. A glutamatergic neuron-specific protein, vesicular glutamate transporter (VGLUT), plays crucial roles in glutamatergic neurotransmission. *Ciona* has one VGLUT gene, *Ci-VGLUT*, which is expressed in the CNS and PNS (Fig. 10.5d, e). Most sensory neurons, including papillar neurons, epidermal neurons, the otolith cell, and ocellus photoreceptor cells are glutamatergic. Namely, the ascidian otolith (a gravity sensor) and photoreceptors use glutamate as a transmitter, as do their vertebrate counterparts, the ear and eye. Axons from rostral epidermal neurons, ocellus photoreceptor cells, and neurons underlying the otolith terminate in the posterior brain. In the posterior brain, a group of glutamatergic interneurons are present, with axons extending to the motor ganglion, suggesting that these neurons control the activity of the motor system. Thus, glutamatergic neurotransmission may play a major role in sensory systems and in the integration of the sensory inputs of the ascidian larva.

Glutamate activates ionotropic glutamate receptors for fast excitatory neurotransmission and activates metabotropic glutamate receptors for slower modulatory effects on transmission. Comprehensive genomic analysis revealed that *Ciona* has 11 putative ionotropic glutamate receptors.

Cholinergic neurons. Cholinergic neurons release acetylcholine (Ach) as a neurotransmitter. Ach, a major excitatory neurotransmitter, is used as a neurotransmitter at the vertebrate neuromuscular junction, in the brain, and in the autonomic nervous system. Ach is produced by choline acetyltransferase (ChAT) and is packed into the synaptic vesicle by vesicular Ach transporter (VACHT). Interestingly, ChAT and VACHT share a common gene locus and regulatory elements for gene transcription, which is conserved among species and is known as the "cholinergic locus." The *Ciona* genome contains the cholinergic locus, and both *Ci-ChAT* and *Ci-VACHT* exhibit similar expression patterns (Fig. 10.5f, g for *Ci-VACHT*).

A few cholinergic neurons located in the posterior brain extend their axons to the motor ganglion. Five pairs of cholinergic neurons are located in the motor ganglion and extend their axons posteriorly (Fig. 10.5f, g). These correspond to

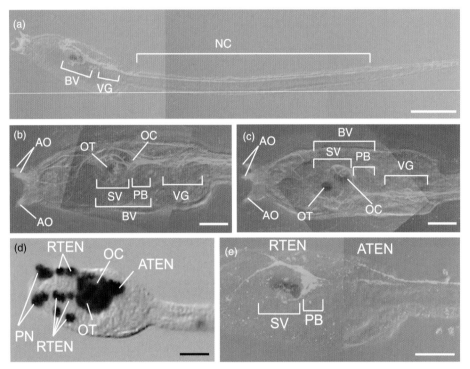

FIGURE 10.5 Neural networks of *Ciona intestinalis* larvae. (a–c) Axons and nerve terminals are visualized by immunofluorescent staining with an antisynaptotagmin antibody. (a) Whole larva. (b) and (c) depict the lateral and dorsal views of the trunk region, respectively. AO, adhesive organ; ATEN, apical trunk epidermal neurons; BV, brain vesicle; NC, nerve cord; OC, ocellus; OT, otolith; PB, posterior brain; PN, papillar neurons; RTEN, rostral trunk epidermal neurons; SV, sensory vesicle; VG, visceral ganglion. Bars represent 50 μm in (a) and 20 μm in (b–i). (d–i) Distribution of neurotransmitter-specific neurons. (d, e) Glutamatergic neurons, visualized by *in situ* hybridization of *Ci-VGLUT* (d), and by immunofluorescent staining with an anti-Ci-VGLUT antibody (e). Glutamatergic neurons are distributed in the central and peripheral nervous systems. (f, g) Cholinergic neurons, visualized by *in situ* hybridization of *Ci-VACHT* (f), and by immunofluorescent staining with an anti-Ci-VACHT antibody (g). Cholinergic neurons are located in PB and VG. Five pairs of cholinergic neurons are located in the visceral ganglion (white arrows). The neurons extend their axons posteriorly and seem to make synapses (white arrowheads) with muscle cells in the anterior part of the tail. (h, i) γ-aminobutyric acid (GABA)/glycinergic neurons, visualized by *in situ* hybridization of *Ci-VGAT* (h), and by immunofluorescent staining with an anti-Ci-VGAT antibody (i). In the anterior nerve cord, the GABA/glycinergic interneurons (arrows) extend their axons anterolaterally (arrowheads). Adapted from Horie *et al.* (2009) *Dev Growth Differ*, **51**, 207–220.)

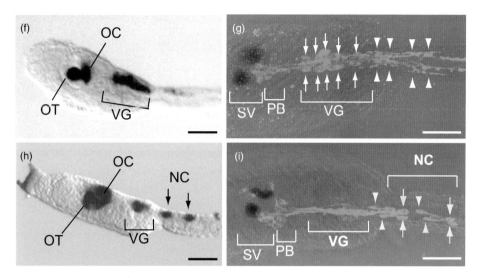

FIGURE 10.5   (*Continued*)

the paired motor nerve tracts in the caudal nerve cord. In the most anterior tail region, several branched axons project ventrolaterally to form nerve terminals on the muscle surface (Fig. 10.5f, g), indicating that the motor neurons of the ascidian larva are cholinergic like their vertebrate counterparts. The *Ciona* genome contains eight genes encoding nicotinic Ach receptors, including those closely related to the vertebrate skeletal muscle nicotinic receptor. However, the expression patterns of these putative Ach receptors remain to be investigated.

GABA/glycinergic neurons. γ-Aminobutyric acid (GABA) is a major inhibitory neurotransmitter in the CNS of both vertebrates and invertebrates, whereas glycine is a vertebrate-specific inhibitory neurotransmitter in the spinal cord. In vertebrates, GABA is mainly used in the brain, and glycine is mainly used in the spinal cord and brainstem. As the vesicular inhibitory GABA/glycine transporter (VGAT) loads both GABA and glycine into the synaptic vesicle, VGAT is used as a GABA/glycinergic neuron marker. The *Ciona* genome has only one VGAT gene, *Ci-VGAT*, which is expressed in the adhesive organ, sensory vesicle, posterior brain, motor ganglion, and dorsal tail region (Fig. 10.5h, i). The GABA/glycinergic neurons form a neural network in the ascidian larva (Fig. 10.5h, i). A few Ci-VGAT-positive cells in the sensory vesicle send axons to posterior brain. In the posterior brain, a number of Ci-VGAT-positive cells form clusters, and most of them send axons to the motor ganglion. In the motor ganglion, Ci-VGAT-positive cells are located dorsal to the cholinergic motor neurons in the central region with respect to the antero-posterior axis. The Ci-VGAT-positive inhibitory neurons in the posterior brain and the motor ganglion may be involved in sensory information processing and/or motor activity control. In the tail, two pairs of VGAT-positive neurons are located in the anterior nerve cord (Fig. 10.5h, i).

The *Ciona* genome contains eight genes encoding putative ionotropic GABA/glycine receptors—seven for GABA and one for glycine. Two genes encoding putative GABA$_B$ receptors, G protein-coupled metabotropic receptors, are reported in *Ciona*. Ci-GABA$_B$ *R1* is expressed in photoreceptor cells, coronet cells, some neurons in the posterior brain, and motor neurons in the motor ganglion. *Ci-GABA$_B$ R2* is expressed in several neurons in the sensory vesicle and a few cells in the motor ganglion.

10.1.3.6 Terminal Differentiation of Neurons in the Larval CNS
Otp, Meis, Hif, Nkx2.1, Six3/6, FoxB, and *FoxHa* are expressed in the ventral sensory vesicle. As *Otp, Meis*, and *Nkx* are specifically expressed in the vertebrate hypothalamus, the most ventral part of the diencephalon, it appears that the ventral region of the *Ciona* sensory vesicle corresponds to the vertebrate hypothalamus. Brain vesicles were isolated by using transgenic embryos carrying Ci-β-tubulin (promoter)::Kaede, which resulted in robust Kaede expression in the *Ciona* larval CNS. Microarray analyses of the isolated brain vesicles identified ~560 genes that are preferentially expressed in the larval brain, and display region-specific expression profiles. Among these genes, 11 encoded neurohormone peptides, including hypothalamic peptides such as gonadotropin-releasing hormone and oxytocin/vasopressin. Six of the identified peptide genes are novel. Genes encoding receptors for some of the peptides are also expressed in the brain. Interestingly, most of the peptide genes are expressed in the ventral brain, suggesting homology of the ventral region of the *Ciona* sensory vesicle with the vertebrate hypothalamus, as mentioned above. However, the mechanisms underlying the terminal differentiation of the larval neurons and ependymal cells remain to be elucidated.

10.2 SENSORY SYSTEMS IN THE LARVAL CNS

The sensory vesicle includes two sensory organs, the otolith and ocellus (Fig. 10.1a and Fig. 10.5a). The otolith senses gravity and the ocellus senses light. The presence of another sensory organ, perhaps one that senses pressure by coronet cells, still remains to be determined. The otolith is composed of a single cell with melanin pigments. The ocellus consists of a pigment cell and three lens cells. Approximately 30 photoreceptor cells surround the ocellus proper. The a10.97 pair forms an equivalent group in the sense that each cell in the pair has the ability to differentiate into either an otolith or an ocellus pigment cell.

Different genes are reported to be involved in *Ciona* and *Halocynthia* with respect to the otolith versus ocellus fate determination of the a10.97 pair. The specification of pigment cells in *Ciona* embryos starts around gastrulation. Between the two daughter cells of a9.49, a10.97 and a10.98, only the a10.97 pair differentiates into pigment cells. FGF signaling is involved in the induction. The a10.97 pair is physically closer to the source of FGF-signaling than the a10.98 pair, leading to the induction of pigment cell fate exclusively in the a10.97 pair. FGF signaling activates the transcription of *Tcf* in the a10.97 pair, enabling the cells to respond to Wnt signaling.

By contrast, the otolith vs. ocellus fate determination in *Halocynthia* embryos is regulated by secreted BMP and its antagonist Chordin, both of which are expressed in the cells surrounding a10.97. BMP promotes the otolith pigment cell fate, whereas Chordin promotes the ocellus pigment cell fate by suppressing the BMP-mediated otolith pigment cell fate.

10.3 PERIPHERAL NERVOUS SYSTEM

Ascidian larvae also contain a peripheral nervous system, containing neurons embedded in the epidermal layer. The adhesive papillae at the most anterior region of the larval trunk contain specialized epidermal and mechanosensory-like neurons (Fig. 10.5d, e). The position and number of the epidermal sensory neurons varies among individuals. In *Ciona*, the epidermal sensory neurons express VGLUT, suggesting that the neurons are glutamatergic in nature. The epidermal neurons extend protrusions into the tunic. As the tunic is the border between the body and the environment, the protrusions of the larval epidermal sensory neurons may enhance the sensitivity of the sensory neurons.

As discussed in Chapter 8, the epidermal sensory neurons in the trunk originate from a-line cells, while those in the tail arise from b-line cells. FGF signaling from the vegetal blastomeres induces the differentiation of the epidermal sensory neurons. The tail epidermis is subdivided into three regions along the dorsal–ventral axis—the dorsal midline, ventral midline, and lateral region. The tail epidermal neurons are formed along the dorsal and ventral midline. The signaling molecules of FGF9/16/20 and ADMP (antidorsalizing morphogenetic protein) are responsible for the regionalization of the epidermis by inducing the midline fate. The paired tail epidermal neurons are well separated from each other, suggesting the presence of a mechanism that restricts the number of epidermal cells. Indeed, lateral inhibition induced by Notch-Delta signaling has been shown to limit the number of epidermal neurons.

10.4 REMODELING OF THE CENTRAL NERVOUS SYSTEM DURING METAMORPHOSIS

A major role of the ascidian larval CNS is the control of swimming behavior by innervating the tail muscle. During metamorphosis, the tail is lost and adult organs such as the heart, endostyle, gut, and gill are formed. Therefore, ascidians have to remodel the CNS during metamorphosis in order to change the organs that are regulated. The neurohypophysial duct forms a connection between the most anterior larval sensory vesicle and the oral siphon primordium or stomodeum. It has been thought that the larval CNS degenerates during metamorphosis and that the adult CNS is newly formed from cells consisting of the wall of the neurohypophysial duct. However, recent evidence supports the continuity of larval CNS cells to adult CNS cells.

First, *Phox2* is specifically expressed in the *Ciona* larval CNS neck region, which contains nonneuronal ependymal cells. Labeling these cells, by putting yellow fluorescent protein (YFP) under control of the *cis*-element of *Phox2*, demonstrates that these cells become neurons during metamorphosis and extend cholinergic axons to muscles associated with the branchial basket in metamorphosed juveniles. These neurons thus are potentially homologous to vertebrate cranial motoneurons of the branchiovisceral class.

Second, more sophisticated tracing of larval CNS cells using the photoconvertible fluorescent protein Kaede shows that cells constituting the major parts of the larval CNS—including the sensory vesicle, neck, and visceral ganglion—remain after metamorphosis and are recruited during the formation of the adult CNS (Fig. 10.6). By contrast, the cells in the tail nerve cord disappear after metamorphosis (Fig. 10.6a). Cell type-specific tracing using Kaede shows that most of the larval neurons disappear during metamorphosis. However, the cholinergic neurons at the motor ganglion are exceptions; a few cholinergic neurons remain after metamorphosis and are integrated into the adult CNS (Fig. 10.6b). On the other hand, many larval ependymal cells are inherited into the adult CNS and become the source of adult neurons (Fig. 10.6a). Therefore, in *Ciona*, the basic scaffold of the CNS is transferred from larva to adult while the neurons and the neuronal

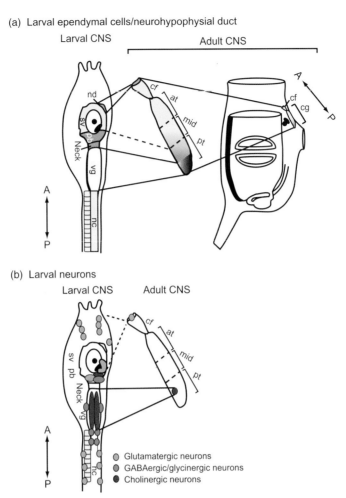

FIGURE 10.6   The relationship between the larval and adult CNS with respect to larval ependymal cells (a) and larval neurons (b); nd, neurohypophysial duct; sv, sensory vesicle; pb, posterior brain; vg, visceral ganglion; nc, nerve cord; cf, ciliated funnel; at, mid and pt, anterior, middle and posterior parts of the cerebral ganglion, respectively; cg, cerebral ganglion; A, anterior; P, posterior. (Adapted from Horie *et al.* (2011) *Nature*, **469**, 525–528.)

network are remodeled to fit the transformed body plan. The glial cells in the *Ciona* larval CNS are classified as ependymal cells; however, they may not be uniform cells. Some of the ependymal cells may have the stem cell- or neuroblast-like ability to produce neurons. Further characterization of cells that produce adult neurons will be important in determining the mechanisms of adult CNS formation.

10.5   THE ADULT NERVOUS SYSTEM

The adult neural complex which is located between the oral and atrial siphons comprises the CNS and the neural gland (Fig. 1.1b and Fig. 10.7a). The former consists of the cerebral ganglion, the nerves of the body wall, the visceral nerve, the dorsal strand plexus and sensory structures connected to the siphons (Fig. 10.7a). The cerebral ganglion includes most of the neural cell bodies of CNS and nerves leaving from the ganglion innervate the siphons, body wall, and caudal viscera (Fig. 4.3f). The latter neural gland connects to the pharyngeal lumen via the ciliated duct and the ciliated funnel (partially filled by the dorsal tubercle). The neural gland and ciliated duct are likely homologous to the circumventricular organs (e.g., the choroid plexus) of vertebrates, with a putative role in controlling the homeostasis of the fluid surrounding the neural complex.

Although genes involved in the development and function of the adult CNS still remains to be elucidated, the ascidian adult CNS provides an appropriate experimental system to study cellular and molecular mechanisms involved in regeneration of the nervous system. Namely, the adult *Ciona* CNS has a capacity to not only regenerate

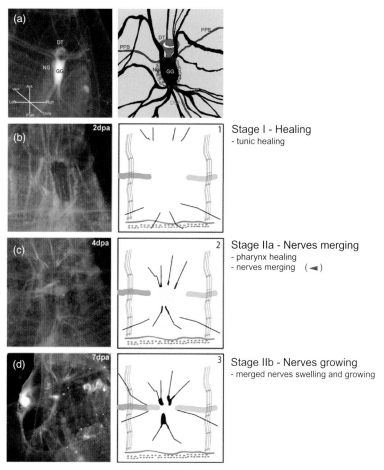

Stage I - Healing
- tunic healing

Stage IIa - Nerves merging
- pharynx healing
- nerves merging (◄)

Stage IIb - Nerves growing
- merged nerves swelling and growing

FIGURE 10.7   Adult *Ciona* CNS and its capacity of regeneration. (a) A neural complex composed of the neural ganglion (GG), the neural gland (NG) and the dorsal tubercle (DT) (left). Nomenclature of the ganglion; peripharyngeal band (PPB), left anterior nerve (LA), right anterior nerve (RA), left posterior nerve (LP), right posterior nerve (RP), dorsal strand plexus (DSP) (right). (b–g) The process and the stages of regeneration. The first column shows the various stages of the regeneration process. The second column provides a short description of the main features of the different stages of the process. (Adapted from Dahlberg *et al.* (2009) *PLos ONE*, **4**, e4458.)

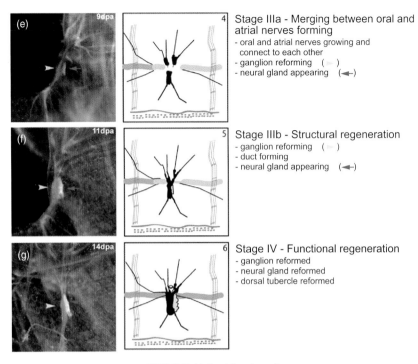

FIGURE 10.7 *(Continued)*

nerves but also form a complete system, which is quite unusual in the chordate phylum. As shown in Fig. 10.7, the regeneration process is subdivided into four stages. New neural structures are derived from the anterior and posterior nerve endings. A blastemal structure is implicated in the formation of new neural cells. Although the rate of regeneration is negatively correlated with adult size, the regeneration is completely functional, demonstrating *Ciona* is as a useful system for studies on regeneration of the brain, brain-associated organs and nerves.

SELECTED REFERENCES

BERTRAND, V., HUDSON, C., CAILLOL, D., POPOVICI, C. *et al.* (2003) Neural tissue in ascidian embryos is induced by FGF9/16/20, acting via a combination of maternal GATA and Ets transcription factors. *Cell*, **115**, 615–627.

CANESTRO, C., BASSHAM, S., POSTLETHWAIT, J. (2005) Development of the central nervous system in the larvacean *Oikopleura dioica* and the evolution of the chordate brain. *Dev Biol*, **285**, 298–315.

DAHLBERG, C., AUGER, H., DUPONT, S., SASAKURA, Y. *et al.* (2009). Refining the *Ciona intestinalis* model of central nervous system regeneration. *PLos One*, **4**, e4458.

DUFOUR, H.D., CHETTOUH, Z., DEYTS, C., DE ROSA, R. *et al.* (2006) Precraniate origin of cranial motoneurons. *Proc Natl Acad Sci USA*, **103**, 8727–8732.

HAMADA, M., SHIMOZONO, N., OHTA, N., SATOU, Y. *et al.* (2011) Expression of neuropeptide- and hormone-encoding genes in the *Ciona intestinalis* larval brain. *Dev Biol*, **15**, 202–214.

HORIE, T., NAKAGAWA, M., SASAKURA, Y., KUSAKABE, T. (2009) Cell type and function of neurons in the ascidian nervous system. *Dev Growth Differ*, **51**, 207–220.

HORIE, T., KUSAKABE, T., TSUDA, M. (2008) Glutamatergic networks in the *Ciona intestinalis* larva. *J Comp Neurol*, **508**, 249–263.

HORIE, T., SHINKI, R., OGURA, Y., KUSAKABE, T.G. *et al.* (2011) Ependymal cells of chordate larvae are stem-like cells that form the adult nervous system. *Nature*, **469**, 525–528.

HUDSON, C., YASUO, H. (2005) Patterning across the ascidian neural plate by lateral Nodal signalling sources. *Development*, **132**, 1199–1210.

HUDSON, C., LOTITO, S., YASUO, H. (2007) Sequential and combinatorial inputs from Nodal, Delta2/Notch and FGF/MEK/ERK signalling pathways establish a grid-like organisation of distinct cell identities in the ascidian neural plate. *Development*, **134**, 3527–3537.

IMAI, J.H., MEINERTZHAGEN, I.A. (2007a) Neurons of the ascidian larval nervous system in *Ciona intestinalis*: I. Central nervous system. *J Comp Neurol*, **501**, 316–334.

IMAI, J.H., MEINERTZHAGEN, I.A. (2007b) Neurons of the ascidian larval nervous system in *Ciona intestinalis*: II. Peripheral nervous system. *J Comp Neurol*, **501**, 335–352.

IMAI, K.S., STOLFI, A., LEVINE, M., SATOU Y (2009) Gene regulatory networks underlying the compartmentalization of the *Ciona* central nervous system. *Development*, **136**, 285–293.

MEINERTZHAGEN, I.A., OKAMURA, Y. (2001) The larval ascidian nervous system: the chordate brain from its small beginnings. *Trends Neurosci*, **24**, 401–410.

MEINERTZHAGEN, I.A., LEMAIRE, P., OKAMURA, Y. (2004) The neurobiology of the ascidian tadpole larva: recent developments in an ancient chordate. *Annu Rev Neurosci*, **27**, 453–485.

MORET, F., CHRISTIAEN, L., DEYTS, C., BLIN, M. *et al.* (2005) Regulatory gene expressions in the ascidian ventral sensory vesicle: evolutionary relationships with the vertebrate hypothalamus. *Dev Biol*, **277**, 567–579.

NICOL, D., MEINERTZHAGEN, I.A. (1991) Cell counts and maps in the larval central nervous system of the ascidian *Ciona intestinalis* (L.). *J Comp Neurol*, **309**, 415–429.

NISHIDA, H., SATOH, N. (1989) Determination and regulation in the pigment cell lineage of the ascidian embryo. *Dev Biol*, **132**, 355–367.

NISHINO, A., BABA, S.A., OKAMURA, Y. (2011) A mechanism for graded motor control encoded in the channel properties of the muscle ACh receptor. *Proc Natl Acad Sci USA*, **108**, 2599–2604.

OGURA, Y., SAKAUE-SAWANO, A., NAKAGAWA, M., SATOH, N. *et al.* (2011) Coordination of mitosis and morphogenesis: role of a prolonged G2 phase during chordate neurulation. *Development*, **138**, 577–587.

OKAMURA, Y., NISHINO, A., MURATA, Y., NAKAJO, K. *et al.* (2005) Comprehensive analysis of the ascidian genome reveals novel insights into the molecular evolution of ion channel genes. *Physiol Genomics*, **22**, 269–282.

SQUARZONI, P., PARVEEN, F., ZANETTI, L., RISTORATORE, F. *et al.* (2011) FGF/MAPK/Ets signaling renders pigment cell precursors competent to respond to Wnt signal by directly controlling *Ci-Tcf* transcription. *Development*, **138**, 1421–1432.

SASAKURA, Y., MITA, K., OGURA, Y., HORIE, T. (2012) Ascidians as excellent chordate models for studying the development of the nervous system during embryogenesis and metamorphosis. *Dev Growth Differ* **54**, 420–437.

STOLFI, A., LEVINE, M. (2011) Neuronal subtype specification in the spinal cord of a protovertebrate. *Development*, **138**, 995–1004.

TAKAMURA, K. (1998) Nervous network in larvae of the ascidian *Ciona intestinalis*. *Dev Genes Evol*, **208**, 1–8.

WADA, H., SAIGA, H., SATOH, N., HOLLAND, P.W.H. (1998) Tripartite organization of the ancestral chordate brain and the antiquity of placodes: insights from ascidian *Pax-2/5/8, Hox* and *Otx* genes. *Development*, **125**, 1113–1122.

YOSHIDA, K., SAIGA, H. (2011) Repression of Rx gene on the left side of the sensory vesicle by Nodal signaling is crucial for right-sided formation of the ocellus photoreceptor in the development of *Ciona intestinalis*. *Dev Biol*, **354**, 144–150.

CHAPTER 11

MESENCHYME

It was previously thought that the so-called mesenchymal cells of the ascidian embryo give rise to most of the adult mesodermal tissues and organs. However, recent lineage tracing experiments have revealed that the embryonic mesenchymal cells only give rise to tunic cells and blood (coelomic) cells. By contrast, two other mesenchymal lines, namely the TLCs and TVCs, are responsible for the development of major adult mesodermal tissues. The lineage and specification mechanisms of each of the three lines have been determined at the single cell level.

11.1 EMBRYONIC LINEAGES AND THEIR ADULT PROGENY

11.1.1 Mesenchyme Cells

The so-called mesenchymal cells originate from the B-line. The mesenchymal potential is passed through B5.1 and B5.2 of the 16-cell embryo, B6.2 and B6.4 of the 32-cell embryo, and B7.3 and B7.7 of the 64-cell embryo (Fig. 2.2 and Fig. 2.3). B7.7 is restricted to mesenchymal cells (Fig. 11.1a). B7.3 has the potential to form mesenchyme and notochord. During the next division to form the 112-cell embryo, B8.5 is restricted to mesenchyme (Fig. 11.1a) while B8.6 becomes only secondary notochord (see Chapter 9).

Lineage tracing experiments show that the mesenchymal cells of the *Halocynthia* embryo give rise only to tunic cells (Fig. 11.1c), while those of the *Ciona* embryo give rise to tunic and blood cells. In *Ciona*, there are no differences between B8.5- and B7.7-derived mesenchymal cells, in terms of the distribution and morphology of the tunic cells. However, as discussed below, B8.5 and B7.7 do not always have identical properties in term of the specific genes that they express in the cells during embryogenesis.

11.1.2 Trunk Lateral Cells

Trunk lateral cells (TLCs) are exclusively derived from the A7.6 pair of the 64-cell embryo, via A5.2 of the 16-cell, and A6.3 of the 32-cell embryo (Fig. 11.1a, b). Lineage tracing studies indicate that, in *Halocynthia*, TLCs give rise to blood cells, body-wall muscle (oral siphon and longitudinal mantle muscle), and epithelium of the first/second gill slits of the adult (Fig. 11.1b, c). However, in *Ciona*, TLCs also give rise to oral siphon muscle, epithelium of the first/second gill slits, and blood cells, but do not contribute to

Developmental Genomics of Ascidians, First Edition. Noriyuki Satoh.
© 2014 John Wiley & Sons, Inc. Published 2014 by John Wiley & Sons, Inc.

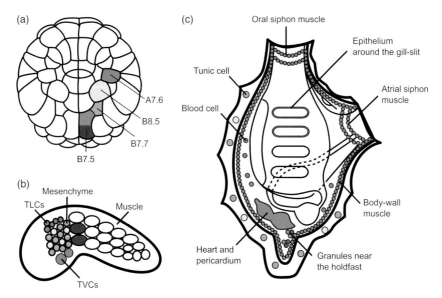

FIGURE 11.1 Diagram illustrating the mesenchymal cell lineage. (a) Vegetal view of the 112-cell embryo showing four pairs of blastomeres that are destined to give rise to adult mesodermal tissues. A7.6 is a trunk lateral cell (TLC) line, B8.5 and B7.7 comprise the so-called mesenchymal cell line, and B7.5 is the trunk ventral cell (TVC) line. (b) Sagittal section showing the position of mesenchyme, TLCs, TVCs, and muscle. (c) Diagram illustrating the anatomy of a *Halocynthia* juvenile. Lateral view, dorsal is to the right; the holdfast is to the bottom. In all figures, the colors of blastomeres, larval tissues, and juvenile tissues show lineage relationships. (Adapted from Hirano and Nishida (1997) *Dev Biol*, **192**, 199–210.)

longitudinal mantle muscle. Instead, the tunic cells and stomach are derived from TLCs. Specifically, the TLCs give rise to a part of the stomach in *Ciona* juveniles, reminiscent of the visceral vertebrate mesoderm. Accordingly, there may be a conserved mechanism underlying stomach formation within chordates. Another possibility is that TLCs have intrinsic endodermal as well as mesodermal cell properties, because they display alkaline phosphatase enzymatic activity, which is a marker of ascidian embryonic endoderm cells. The differences observed between *Halocynthia* and *Ciona* embryos may be due to species-specific properties, as discussed in Chapter 17.

11.1.3 Trunk Ventral Cells

Trunk ventral cells (TVCs) originate from the B7.5 pair of the 112-cell embryo (Fig. 11.1a, b). As will be discussed in Chapter 13, when the B7.5 lineage is traced, B7.5 first divides equally into B8.9 and B8.10 in the early gastrulae. Then, in late gastrulae/neurulae, B8.9 and B8.10 divide into smaller anterior (rostral) daughter cells and larger posterior (caudal) daughter cells, respectively. During larval tail extension, the rostral cells separate from their caudal sisters and become TVCs, and the caudal cells become anterior muscle cells of the larval tail (Fig. 11.1b, c). The tail muscle cells remain in their initial position and undergo no further cell divisions or migration. By contrast, TVCs migrate anteriorly along the ventral surface of the trunk endoderm, and then meet along the ventral midline. TVCs give rise to juvenile body-wall muscle (atrial siphon and latitudinal mantle), heart, and pericardium (Fig. 11.1c).

11.2 MOLECULAR CASCADES INVOLVED IN MESENCHYMAL SPECIFICATION

The molecular and cellular mechanisms underlying the development of TVC-derived cardiac cells will be discussed in Chapter 13, with respect to their involvement in juvenile heart formation. Here, we discuss the molecular mechanisms involved in the differentiation of A7.6 (TLC)-, B8.5-, and B7.7-line mesenchymal cells.

In *Ciona* embryos, A7.6, B8.5, and B7.7 each are distinguished by the expression profiles of nine mesenchyme-specific genes. As shown in Fig. 11.2, of these, five genes are expressed in A7.6, B8.5, and B7.7 (A7.6/B8.5/B7.7-specific genes; e.g., *Ci-AKR1a*), but two are expressed only in B8.5 and B7.7 (B8.5/B7.7-specific genes; e.g. *Ci-PER-like*), and another two are expressed only in B7.7 (B7.7-specific genes; e.g., *Ci-WBSCR27-like*). These molecular markers are used in genetic studies for investigating the specification of the three mesenchymal lines. FGF9/16/20 is required for the development of all three lines. The bHLH transcription factor, *Twist-like1*, is associated with two different (TLC and B8.5/B7.7), but partially overlapping molecular specification mechanisms.

11.2.1 Mesenchyme Cells

FGF9/16/20 is downstream of maternal β-catenin and is expressed in endodermal cells of the 16-cell and 32-cell embryo. B6.4 of the 32-cell embryo is a precursor of mesenchymal and muscle cells. At the next cell division, B6.4 divides into B7.7 (mesenchyme) and B7.8 (muscle) (Fig. 11.2b). B7.7 is in contact with endodermal cells and receives FGF signals inducing differentiation to mesenchyme (Fig. 11.2b, f). By contrast, B7.8 does not make contact with endodermal cells and, on receiving input from the maternal factor, Macho-1, gives rise to muscle.

B7.3 of the 64-cell embryo is a precursor of mesenchymal and notochord cells (Fig. 11.2b). At the next cell division, B7.3 divides into B8.5 (mesenchyme) and B8.6 (notochord) (Fig. 11.2c). B8.5 receives FGF signals inducing differentiation to mesenchyme (Fig. 11.2e). B8.6 receives a Nodal signal from the neighboring b6.5-line cells that induces the development of secondary notochord cells (see Fig. 9.2b).

The functional suppression of *FGF9/16/20* leads to an absence of mesenchyme formation, and the embryonic cells assume default muscle fate. A bHLH transcription factor gene *Twist-like1* is downstream of *FGF9/16/20* and is expressed in B7.7/B8.5 and A7.7 (TLC) mesenchymal cells (Fig. 11.2d-f). *Twist-like1* is required for the expression of all the mesenchyme-specific genes in both B8.5- and B7.7-line cells. Maternal *macho-1* and zygotic *Otx* is also involved in the *Twist-like1* expression. The overexpression of *FGF9/16/20* or *Twist-like1* upregulates the expression of B7.7/B8.5/A7.6-specific and B7.7/B8.5-specific genes, but downregulates the expression of B7.7-specific genes, suggesting that the B7.7 and B8.5 line-specific genes are under differential control.

11.2.2 Trunk Lateral Cells

By contrast to B8.5/B7.7-line mesenchymal cells, where the FGF9/16/20 signal is sufficient to initiate *Twist-like1* transcription, only early FGF9/16/20 signal is required for

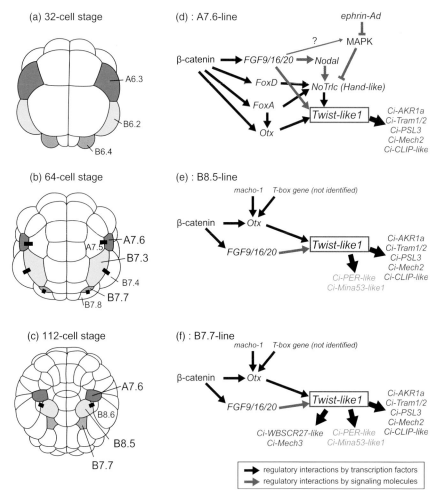

FIGURE 11.2 A model for the genetic cascades underlying mesenchymal cell specification in *Ciona* embryos. (a–c) Fate specification of three mesenchymal cell lines at (a) the 32-, (b) 64-, and (c) 112-cell stages. A7.6-line TLCs are shown in green, B8.5-line mesenchyme in yellow, and B7.7-line mesenchyme in brown. Blastomeres connected with bars are sister blastomeres. (d–f) Outline of the molecular cascades for the leading mesenchyme-specific genes in (d) A7.6-line, (e) B8.5-line, and (f) B7.7-line. (Modified by M. Tokuoka based on Tokuoka *et al.* (2004) *Dev Biol*, **288**, 387–396.)

the specification of A7.6-line (TLC) cells (Fig. 11.2d). MAPK signaling is active in the endoderm descendants of A6.3, but is absent from the mesoderm lineage. Inhibition of MAPK signaling results in expanded expression of mesoderm marker genes and loss of endoderm markers, whereas ectopic MAPK activation produces the opposite phenotype: the transformation of mesoderm into endoderm. Here, an Ephrin signaling molecule, *Ci-ephrin-Ad*, is required to establish asymmetric MAPK signaling in the endomesoderm. Namely, reducing *Ci-ephrin-Ad* activity results in ectopic MAPK signaling and conversion of the mesoderm lineage into endoderm. Conversely, misexpression of *Ci-ephrin-Ad* in the endoderm induces ectopic activation of mesodermal marker genes (Fig. 11.2d).

In addition to FGF9/16/20, the activity of *FoxD* and another bHLH factor, *NoTrlc* (*Hand-like*) are required for the fate determination of TLCs (Fig. 11.2d). *FoxD* is expressed at the 32-cell stage in A6.3 and *NoTrlc* is expressed in A7.6 of the 64-cell embryo. FoxD likely controls *NoTrlc* transcription, which directly or indirectly regulates *Twist-like1*, a regulator of the mesenchyme-specific gene expression (Fig. 11.2d).

SELECTED REFERENCES

HIRANO, T., NISHIDA, H. (1997) Developmental fates of larval tissues after metamorphosis in ascidian *Halocynthia roretzi*. I. Origin of mesodermal tissues of the juvenile. *Dev Biol*, **192**, 199–210.

KOBAYASHI, K., SAWADA, K., YAMAMOTO, H., WADA, S., SAIGA, H. *et al.* (2003) Maternal *macho-1* is an intrinsic factor that makes cell response to the same FGF signal differ between mesenchyme and notochord induction in ascidian embryos. *Development*, **130**, 5179–5190.

SHI, W., LEVINE, M. (2008) Ephrin signaling establishes asymmetric cell fates in an endomesoderm lineage of the *Ciona* embryo. *Development*, **135**, 931–940.

TOKUOKA, M., IMAI, K.S., SATOU, Y., SATOH, H. (2004) Three distinct lineages of mesenchymal cells in *Ciona intestinalis* embryos demonstrated by specific gene expression. *Dev Biol*, **274**, 211–224.

TOKUOKA, M., SATOH, N., SATOU, Y. (2005) A bHLH transcription factor gene, *Twist-like1*, is essential for the formation of mesodermal tissues of *Ciona* juveniles. *Dev Biol*, **288**, 387–396.

CHAPTER 12

MAKING BLUEPRINT OF CHORDATE BODY: DYNAMIC ACTIVITIES OF REGULATORY GENES

Chapters 6–11 have discussed how ascidian embryonic cells are specified to give rise to one type of larval tissues and what kinds of regulatory genes (transcription factors and signaling pathway molecules) are involved in the specification and subsequent differentiation of the cells. This chapter is a challenge to understanding how the blueprint of chordate body plan is established by the embryo as a whole. The *Ciona intestinalis* genome contains ~318 core transcription factors and 110 major signaling pathway molecules (Chapter 3). How is the expression of these genes harmoniously controlled to derive a coordinate function to construct the blueprint?

12.1 THE EMBRYONIC AXES FORMATION

The definition of embryonic axis is somewhat semantic and this is the case in ascidians. In ascidians, dynamic rearrangement of egg cytoplasm after fertilization and unconventional mode of gastrulation make it difficult *per se* to exactly superpose future embryonic axes on the egg and early embryo. As discussed in Chapter 2, the unfertilized egg of ascidians is polarized along the animal–vegetal (An–Vg) axis as polar body formation point represents the animal pole. When this pole is used as the reference point, the anterior–posterior (A–P) axis is established as approximately perpendicular to the An–Vg axis during ooplasmic segregation (Fig. 12.1a). As embryonic cells of the animal hemisphere do not exhibit complicated positional rearrangement, the A–P axis of the egg and early embryo corresponds to the A–P or cranio–caudal axis of larvae (Fig. 12.1a). The second, dorsal–ventral (D–V) axis becomes apparent after gastrulation and runs orthogonal to the A–P axis (Fig. 12.1a). As the vegetal pole becomes covered with dorsal tissues at gastrulation, the An–Vg axis approximately corresponds to the V–D axis. Therefore, when viewed from the vegetal pole, the left–right (L–R) axis of the ascidian embryo appears opposite to that of the *Xenopus* embryo (Fig. 12.1a, b).

Developmental Genomics of Ascidians, First Edition. Noriyuki Satoh.
© 2014 John Wiley & Sons, Inc. Published 2014 by John Wiley & Sons, Inc.

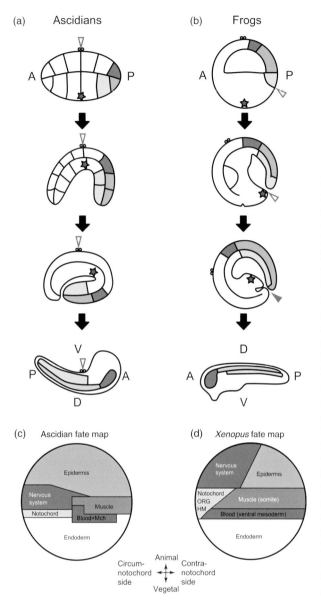

FIGURE 12.1 (a, b) Schematic representation of morphogenetic movements during gastrulation of ascidians (a) and frogs (b). Presumptive notochord region is yellow colored. Anterior and posterior neural tube is colored dark and light green, respectively. Polar bodies indicate the animal pole. Asterisks point the original vegetal pole. The red arrowhead represents the point held in a constant position. Embryonic axes at tailbud stage are indicated at the bottom. (Modified from Nishida (2005) *Dev Dyn*, **233**, 1177–1193.). (c, d) Fate maps of ascidians and *Xenopus* blastulae. Lateral views. Circum-notochord side is to the left. Note the similarity of topography in presumptive tissue territories in the two fate maps. ORG, organizer; HM, head mesoderm; Mch, mesenchyme (precursors of adult tunic cells). (Adapted from Lemaire *et al.* (2008) *Curr Biol*, **18**, R620–R631.)

12.2 THE FATE MAP

During ascidian embryogenesis, anterior–animal blastomeres give rise to head/trunk epidermis of the larva, and posterior–animal blastomeres develop into tail epidermis (Fig. 12.1c). The polar bodies attach to the ventral midline at the boundary of trunk and tail of the tadpole (Fig. 12.1a). Anterior–vegetal blastomeres give rise mainly to endoderm, notochord, and nerve cord. Via gastrulation movement, notochord and nerve cord cells are located in the larval tail. Posterior–vegetal blastomeres develop mainly into endoderm, mesenchyme, and muscle. Muscle cells position in the tail, while endoderm and mesenchyme cells are in the trunk region. On the basis of these developmental schemes, the fate map is drawn as shown in Fig. 12.1c. The ascidian fate map shows a high degree of topographic similarity with the fate map of *Xenopus* (Fig. 12.1d), suggesting an organization conserved between ascidians and vertebrates. In both ascidians and vertebrates, ectoderm is derived from the animal hemisphere, endoderm from the vegetal pole region, and mesoderm from the equator of the embryo. However, each territory differs in the relative proportion of the blastula. In ascidians, the territory of nervous system occupies more equatorial region of the superposed blastula, and the muscle territory positions only the posterior half (Fig. 12.1c, d).

12.3   FROM MATERNAL FACTORS TO ZYGOTIC ACTIVITIES OF REGULATORY GENES

As discussed in previous chapters, three major, maternally provided factors are engaged in the establishment of the basic body plan of an early ascidian embryo (Fig. 12.2). First, β-catenin acts as the maternal endoderm determinants at the vegetal side (Chapter 7). β-catenin activates at least eight zygotic genes including *FoxA, FoxD*, and *FGF9/16/20*, which play essential roles in mesoderm formation (Chapters 9 and 11). Second, on the opposite side, the ectoderm is specified by the action of a maternal transcription factor, GATAa (most similar to vertebrate GATA4/5/6), which functions together with the GATA-cofactor Friend-of-GATA (FOG) (Chapter 8). GATAa mRNA and protein are ubiquitously distributed, and the activity of GATAa is restricted to the animal hemisphere by vegetal β-catenin, although the molecular mechanism for this restriction remains to be elucidated. The third factor is the muscle determinant macho-1, a Gli family zinc-finger transcription factor (Chapter 6). The positioning and segregation of *macho-1* mRNA are closely associated with the PVC. As discussed in Chapter 5, the PVC is a complex multi-layered cortical structure enriched in cER, germ plasm, and localized mRNAs collectively referred to as PEM. By the 8-cell stage, the PVC concentrates into a subcellular structure, the CAB. Inheritance of this structure is responsible for a series of unequal cleavages of most vegetal cells. Macho-1 functions as a muscle determinant via the zygotic activation of *Tbx6*, followed by that of myogenic bHLH factor genes (Chapter 6).

Other maternal factors also support the establishment of the basic body plan (Chapter 5). Wnt5 acts in mesenchyme and muscle fate specification. Pem is required for the proper positioning of the mitotic spindle during the early unequal cleavages. Pem is thought to act as an adaptor protein pulling on the distal tip of astral microtubules to attract the centrosome to the PVC or CAB. The POPK-1 kinase is required for the positioning and concentration of the cER, macho-1 mRNA, germ plasm, and PEM mRNAs to the CAB.

12.4   THE ANTAGONISTIC ACTIONS OF FGF AND EPHRIN PATHWAYS TO FORM GERM LAYERS

At the 16-cell stage, twelve transcription factor genes and six signaling ligand genes are zygotically expressed (Fig. 12.2b). Of the signal ligands, FGF9/16/20 and EphrinAd account for significant cell communication events up to the late 32-cell stage (Fig. 12.3A). *FGF9/16/20*, which is downstream of β-catenin, is transcribed in most cells of the vegetal

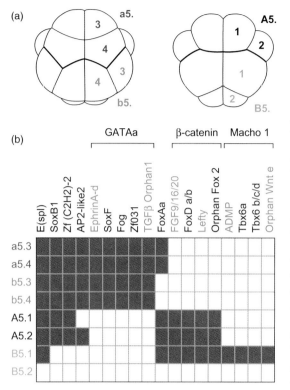

FIGURE 12.2   Onset of zygotic program of regulatory genes in the 16-cell *Ciona* embryo. (a) Animal (left) and vegetal (right) hemispheres of the embryo, on which cells on one side have been labeled (A-line in black, B-line in blue, a-line in red, and b-line in green). (b) Representation of the expression patterns of the 18 regulatory genes zygotically expressed at this stage. Genes in black are transcriptional regulators, genes in green code for signaling ligands. These genes combinatorially define five domains: a-line, b-line, A-line, B5.1, and B5.2 (transcriptionally silent). GATAa, β-catenin, and Macho1 are direct regulators of the indicated coexpressed genes. (Modified from Lemaire (2009) *Dev Biol*, **332**, 48–60.)

hemisphere, except in the transcriptionally silent B5.2 (Fig. 12.2b and Fig. 12.3a). In a complementary manner, *EphrinAd*, which is downstream of GATAa, is transcribed throughout the animal hemisphere during the 16- and 32-cell stages (Fig. 12.2b and Fig. 12.3b). Gain and loss of function for *FGF9/16/20* and *EphrinAd* indicate that they play major, and antagonistic, roles in germ layer specification (Fig. 12.3c–e).

FGF9/16/20 signaling pathway induces notochord, mesenchyme, animal neural tissue, and anterior endoderm (Fig. 6.3A). Suppression of the FGF pathway decreases the number of induced cells. Because all vegetal cells express FGF and all animal cells touch at least one vegetal cell, it is an intriguing question why not all competent cells are induced. The precise mechanisms restricting the response to the inducer in the ectoderm and the mesendoderm differ, but they involve in each case a combination of short-range signaling and precise positioning of the cleavage plane of a bipotential precursor. In the mesenchyme lineage, for example, the contacts established by the bipotential progenitor with FGF-expressing cells are intrinsically polarized and the asymmetric reception of FGF is sufficient to polarize the cell. Polarization of the notochord/nerve cord precursor is more sophisticated as this cell expresses *FGF9/16/20* itself. Its polarization requires a contact with animal cells and involves animally expressed *EphrinAd*. When the function of *EphrinAd* is blocked, polarization of the mother cell is prevented and both daughters form notochord; conversely, if *EphrinAd* is overexpressed, twice as many nerve cord cells form (Fig. 12.3e). At the molecular level, the action of FGF9/16/20 and EphrinAd converge on the MAP kinase ERK, which FGF9/16/20 activates and EphrinAd represses.

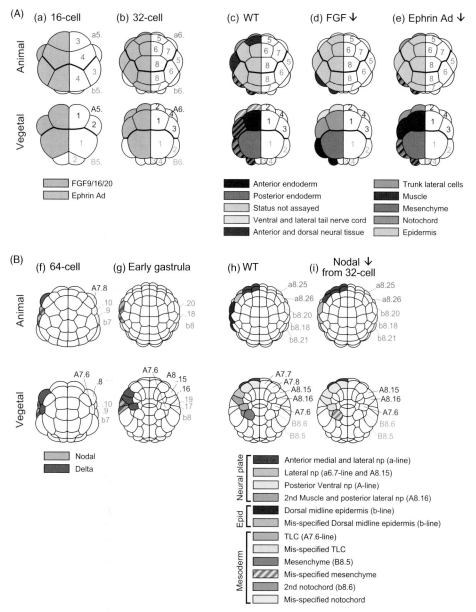

FIGURE 12.3 (A) The FGF/EphrinAd antagonism in the germ layer specification at the 32-cell stage. (a, b) Expression profiles of *FGF9/16/20* (purple) and *EphrinAd* (brown) at (a) the 16- and (b) 32-cells stages. (c–e) Effect on fate specification of interfering between the 16 and late 32-cell stage with FGF or Ephrin signaling. Animal (top) or vegetal (bottom) schemes of embryos on which the left-hand cells have been color-coded for cell fates and the right-hand side indicates cell identities. Fates colored as indicated. (B) Inductions between the 44-cell and early gastrula stages. (f, g) Expression of *Nodal* and *Delta2* at (f) the 64- and (g) early gastrula stages. Expression, normally bilateral, is only represented on the left side of schemes. The identity of cells expressing the genes is indicated on the right-hand side of schemes. (h, i) Effect on cell fate specification of interfering with Nodal and Notch signaling from the 44-cell stage. Cells are color coded according to their fate on the left of each scheme. The names of cells whose fates are affected are indicated on the right. (Modified from Lemaire (2009) *Dev Biol*, **332**, 48–60.)

FGF signaling is also involved in the induction of endoderm (Chapter 7), the dorsal midline tail epidermis (Chapter 8), and neural fate (Chapter 10) during the 32-cell stage. In case of neural induction, the induction occurs after the division of the bipotential epidermis/neural precursors, a5.3 and b5.3. Analysis of 3D digital models of embryos (see Section 12.6) indicates that the cleavage plane of a5.3 and b5.3 partitions the surface of contact with FGF-expressing cells in a very asymmetric manner, and the neural progenitor inherits more than 80% of the contacts with the inducing cells. The contact surface between inducing and responding cells is a critical for the selection of induced cell types. Interestingly, above a certain threshold of contact, cells are fully induced, while no induction occurs below this threshold. As animal cells express *EphrinAd*, the threshold of response to FGF is likely to be defined by the level of extracellular signal-regulated protein kinase (ERK) inhibition provided by EphrinAd in the animal cells.

12.5 THE NODAL AND Delta2 SIGNALING RELAY

From the 44-cell stage onward, Nodal that is activated by FGF9/16/20 in the posterior ectodermal b6.5 plays a major role of cell fate specification of mesenchyme, secondary notochord, and probably TLCs. Nodal acts either directly or via a divergent Notch ligand, Delta2. For example, in the mesenchyme lineage, Nodal is required for proper level of *Twist-like1* expression in B8.5 mesenchyme precursor, and *Twist-like1, Hand*, and *Delta2* in A7.6 TLC precursor (Fig. 12.3B). However, A7.6 marker gene expression does not require direct Nodal signaling from b6.5. As B8.5 precursor is located away from b6.5 precursors, a similar situation may apply in this mesenchyme lineage (Fig. 12.3B).

Nodal also plays a pivotal role in patterning the whole neural plate, and adjacent epidermis, along its medio-lateral axis (Fig.12.3B) (Chapter 10). Between the 32- and 64-cell stages, the Nodal signaling pathway is activated in the lateral precursors of the neural plate (a7.13; A7.8), and confers to these cells their lateral neural plate identity (Fig. 12.3h, i and Fig. 10.3). By the 64-cell stage, Nodal has turned on *Delta2* in b7.9/10 and A7.6 (Fig. 12.3B). Polarization of A7.8 to A8.15 and A8.16 occurs before its cleavage at the 76-cell stage, and this confers its distinct lateral-most identity to A8.16 (Fig. 12.3B). By the early gastrula stage, *Delta2* is expressed in both A8.16 (Column 4) and A8.15 (Column 3) and signals to the next A-line neural cell, A8.8, to which it confers a column 2 identity (Fig. 10.2). Thus, between the 64-cell and early gastrula stage, two temporally separable, short range, Delta2/Notch signaling events confer to the A-line neural precursors their four distinct column identity.

12.6 THE ANISEED DATABASE AS AN INTERFACE FOR ELUCIDATION OF *Ciona* DEVELOPMENTAL PROGRAM

ANISEED is an implementation of a generic digital system, NISEED, to the *Ciona* system. This highly structured data set can be explored via a Developmental Browser, a Genome Browser, and a 3D Virtual Embryo module (Fig. 12.4A). ANISEED includes the detailed anatomical ontologies for these embryos, and quantitative geometrical descriptions of developing cells obtained from reconstructed 3D embryos up to the gastrula stage (Fig. 12.4B). In addition, fully annotated gene model sets are linked to 30,000 high resolution spatial gene expression patterns in wild type and experimentally manipulated conditions and to ~530 experimentally validated *cis*-regulatory regions imported from specialized databases or extracted from 160 literature articles. As discussed above, there are many questions as to how embryonic cells interact each other and how they use regulatory genes to establish the blue print of chordate body plan. ANISEED can provide a system-level understanding of the developmental program through the automatic inference of gene regulatory interactions, the identification of inducing signals, and the discovery and explanation of novel asymmetric divisions.

12.7 STRATEGY OF BODY PLAN FORMATION IN ASCIDIANS

Description and discussion of ascidian embryogenesis as a whole highlight a characteristic strategy adopted by ascidians to form their basic larval body plan. First, three maternal factors are essential as a starting point for the body-plan formation. Second, the maternal information is relayed on zygotic activity of regulatory genes that play pivotal roles in the global germ layer formation. Third, cell–cell communication plays pivotal roles in harmonious shaping up of embryonic regionalization, and fourth, just only four or five signaling pathways act, at very short range, to fine tune cell fate specification dependent on the relative spatial position of embryonic cells.

12.7.1 Invariant Lineage and Mosaic Development

Traditionally, the ascidian embryogenesis has been considered to be driven by invariant cell lineage and a cell-autonomous "mosaic" mode of development. This may be correct in a sense because, without the localized three major maternal factors, the first and basic triggers of development cannot be executed successfully. Furthermore, the ectoderm, primary muscle, and endoderm, totaling 72/112 cells at the onset of gastrulation, are cell-autonomously specified. Finally, as is discussed in Chapter 13, juveniles obtained from *Mesp* loss-of-function morphants cannot form the heart, but the other organs and tissues develop normally in such juveniles.

However, in another sense, cell–cell communication plays a more important role during ascidian early embryogenesis than previously thought. By the mid-gastrula stage, only around 10/300 cells form completely cell-autonomously: the primary muscle, lateral tail epidermis, and possibly some lateral trunk epidermal cells may also form autonomously. In most cases, the inductions occur at very short ranges between contacting cells. Some inductions, like that of the a6.5 neural progenitor by FGF9/16/20 even require a precise partitioning of the surface of contact with inducing cells between induced and uninduced sisters. Future studies with 3D models of embryos should test whether the precise surface of contact is used generally to select induced cells among a competent domain, or whether this is only critical for early FGF signaling. In any case, the high prevalence of short-range signaling probably imposes constraints on the lineage and may explain why it has been so well conserved between distantly related ascidian species.

12.7.2 A Few Signaling Pathways

As discussed above, comparatively few signaling pathways are involved in early fate decisions. FGF, Ephrin, Nodal, and Notch/Delta are the major signals. FGF, Nodal,

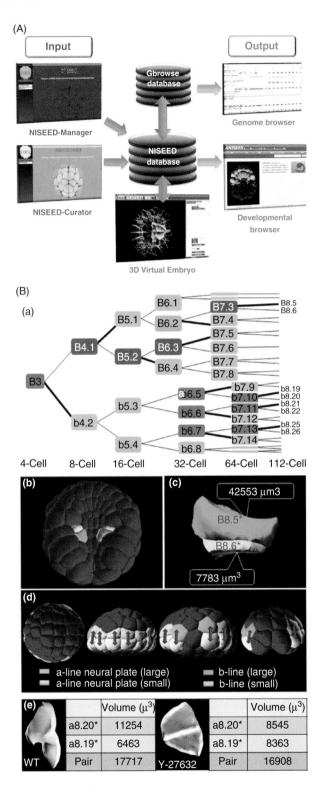

(A)

Input

NISEED-Manager

NISEED-Curator

Gbrowse database

NISEED database

3D Virtual Embryo

Output

Genome browser

Developmental browser

(B)

(a)

B3

B4.1

b4.2

B5.1

B5.2

b5.3

b5.4

B6.1

B6.2

B6.3

B6.4

b6.5

b6.6

b6.7

b6.8

B7.3

B7.4

B7.5

B7.6

B7.7

B7.8

b7.9

b7.10

b7.11

b7.12

b7.13

b7.14

B8.5
B8.6

b8.19
b8.20
b8.21
b8.22

b8.25
b8.26

4-Cell   8-Cell   16-Cell   32-Cell   64-Cell   112-Cell

(b)

(c)   42553 μm3

B8.5*

B8.6*

7783 μm³

(d)

■ a-line neural plate (large)   ■ b-line (large)
■ a-line neural plate (small)   ■ b-line (small)

(e)		Volume (μ³)			Volume (μ³)
	a8.20*	11254		a8.20*	8545
	a8.19*	6463		a8.19*	8363
WT	Pair	17717	Y-27632	Pair	16908

Notch/Delta, and BMP are "classical" ligands implicated in numerous cell fate decisions across evolution. This is the case of the role of Ephrin as an antagonist of FGF signaling. As comparatively few pathways are involved in early fate decisions, each of them can induce several fates in different cells. This requires sophisticated mechanisms to control the competence of cells to respond to inducers. For instance, FGF induces the fate in various cells by between the 16- and late 32-cell stage. This is accomplished by a combinatorial code of ubiquitous maternal ETS factor ETS1/2 and a limited number of distinct regionally active transcription factors in each lineage. In the notochord, ETS1/2, FoxAa, and ZicL activate *Brachyury*. In the animal neural lineages, ETS1/2 and GATAa activate *Otx*. Likewise, in the B7.5 lineage, *Mesp* activation is controlled by FGF and by the combination of two transcription factors, Lhx and Tbx6, whose genes are activated by β-catenin and Macho1, respectively, and then define a domain of competence to respond to FGF during the late gastrula and neurula stages (see Chapter 13). In these cases, the control of the domains of expression/activity of the competence factors occurs cell-autonomously. The fixed ascidian lineage thus appears to regulate cell communication in two complementary ways: it precisely partitions maternal information to define initial competence domains, and it controls the contacts between these competent blastomeres and the short-range signaling sources. A reason why a very few signaling pathways are only used may be their strategy of precocious development. Instead of using many other signaling pathways, the ascidian embryo has evolved by adopting a few pathways that act under a sophisticated tuning-up manner.

12.7.3 Conservation with Vertebrates

The emerging detailed understanding of the ascidian developmental program provides clue to discuss its relationship with vertebrate embryogenesis. Ascidian and vertebrate embryos have very similar fate maps from the onset of gastrulation (Fig. 12.1c, d) and both make extensive usage of inductive processes. However, in the details, inducers and their targets often differ in the two taxa and our current understanding of the ascidian and vertebrate programs points to only a few conserved islands (e.g., the heart specification network; Chapter 13) in a sea of differences. Interestingly, the similarity

---

FIGURE 12.4   The ANISEED database as a digital representation, formalization, and elucidation of early *Ciona intestinalis* developmental program. (A) Overview of the architecture of the NISEED system. (Arrows) Direction of information flow. (B) Systematic identification of unequal cell cleavages: (a) Lineage tree for B3 between the 4- and 112-cell stages. The only blastomeres named are in lineages where unequal divisions occur. (Gray) Cells that divide symmetrically, (light pink) cells with weak asymmetry (index > 15%), (pink) cells with marked asymmetry (index > 25%), (red) cells with strong asymmetry (>50%). (b, c) Example of asymmetry between the 64- and 112-cell stages in the vegetal hemisphere. The yellow (B8.6) and green (B8.5) cells are daughters of the same mother cell (B7.3). Note the important difference in cell volumes between the two sisters. (d) Position of unequally cleaving animal cells between the 76- and 112-cell stages. From the left: animal view, anterior is to the left; frontal view, animal is to the top; lateral view, anterior is to the left, animal to the top; posterior view, animal is to the top. (Blue arrows) Sister cells are linked. (Colors) Lineage and size of cells. (e) Effect of the inhibition of endoderm invagination with the Rho-kinase inhibitor Y-27632. Side views of the a8.19/a8.20 cell pair in wild-type and Y-treated conditions are shown, as well as a measure of the mean volumes of these cells in the two embryos analyzed. (Adapted from Tassy *et al.* (2010) *Genome Res*, **20**, 1459–1468.)

of ascidian and vertebrate programs does not increase significantly as development proceeds and the body forms become increasingly alike. A large-scale comparison of WISH patterns of *Ciona* and zebrafish orthologous genes reveals that a surprising level of divergence of individual expression patterns of both regulatory and nonregulatory genes extends throughout the chordate body-plan formation.

12.8   GENE REGULATORY NETWORK IN EARLY *Ciona* EMBRYOS

Recent progress has been remarkable in identifying the regulatory genes and signaling pathways responsible for the development of a variety of tissues and organs in various metazoans. This information has been integrated to produce gene regulatory networks (GRNs) embodying the functional interconnections among the genes responsible for a given developmental process. The success has been obtained for the specification of endomesoderm in the pregastrula sea urchin embryo and the dorsal–ventral patterning of the early *Drosophila* embryo. As discussed above, *C. intestinalis* also provides an ideal experimental system to elucidate GRNs underlying specification of embryonic cells.

12.8.1   A Sub-Complete Set of Description of Zygotic Expression of Regulatory Genes

Ciona embryogenesis has been elucidated through the deciphering of the patterns of activities of ~318 core sequence-specific TF genes and 110 major SPM genes. Comprehensive WISH assays show that, of these regulatory genes, 65 TFs and 26 SPMs are zygotically expressed between the 16-cell and early gastrula stages of embryogenesis; the expression of 53 TF genes and 23 SPM genes is exclusively zygotic.

Every blastomere from the 16-cell to early gastrula (around 112-cell) stage can be assigned a unique identity on the basis of the expression of specific combinations of TF genes (regulatory code) (Fig. 12.5A, B). There is a close correspondence between establishing different regulatory codes and forming diverse cell lineages. For example, the blastomeres that form the primitive gut (endoderm; A6.1, B6.1, and A6.3 at the 32-cell stage; and A7.1, A7.2, A7.5, B7.1, and B7.2 in 64-cell embryos) have slightly different regulatory codes during early cleaving embryos but acquire identical codes at the early gastrula stage (Fig. 12.5A, B). All lineages except for the B7.5 blastomeres, which form the heart (trunk ventral cells) and anterior tail muscles, achieve clonal restriction before gastrulation. Thus, the hierarchical clustering of cell identities with similar regulatory codes accurately reflects cell lineages and the clonal restriction of cell fate and illuminates at what point key molecular interactions occur to establish a unique identity for each cell.

12.8.2   Provisional Gene Regulatory Networks

The provisional GRN has been achieved by functional assays of 79 TF and 25 SPM genes (Fig. 12.6a). Overall, the resulting analysis provides more than 3000 combinations of gene expression profiles in morphant backgrounds (Fig. 12.6a). In addition, the GRN has been assigned to each of embryonic cells. Therefore, such information is used to create a provisional circuit diagram showing the interconnections among the genes controlling cell fate specification and the initial phases of tissue differentiation (Fig. 12.6a). This large

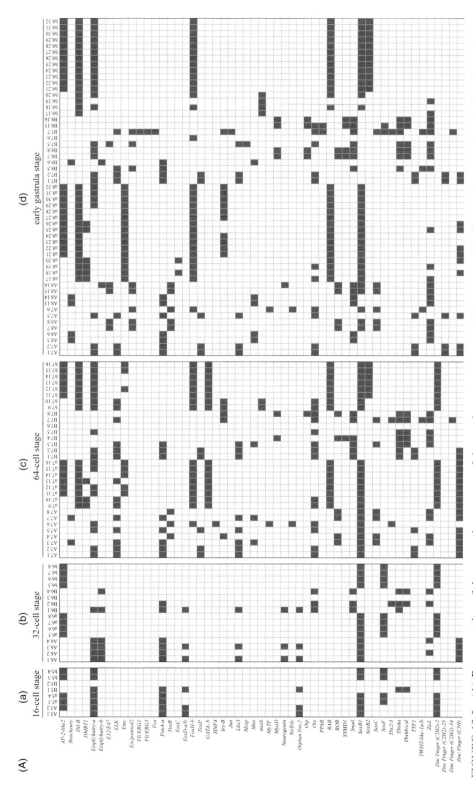

FIGURE 12.5 (A) Representation of the expression patterns of the regulatory genes zygotically expressed in *Ciona* embryos at the (a) 16-cell, (b) 32-cell, (c) 64-cell, and (d) early gastrula stages. (Modified from Imai *et al.* (2006) *Science,* **312,** 1183–1187.) (B) Representation of the expression patterns of the zygotically expressed regulatory genes in each blastomere at the (e) 16-cell, (f) 32-cell, and (g) 64-cell stages.

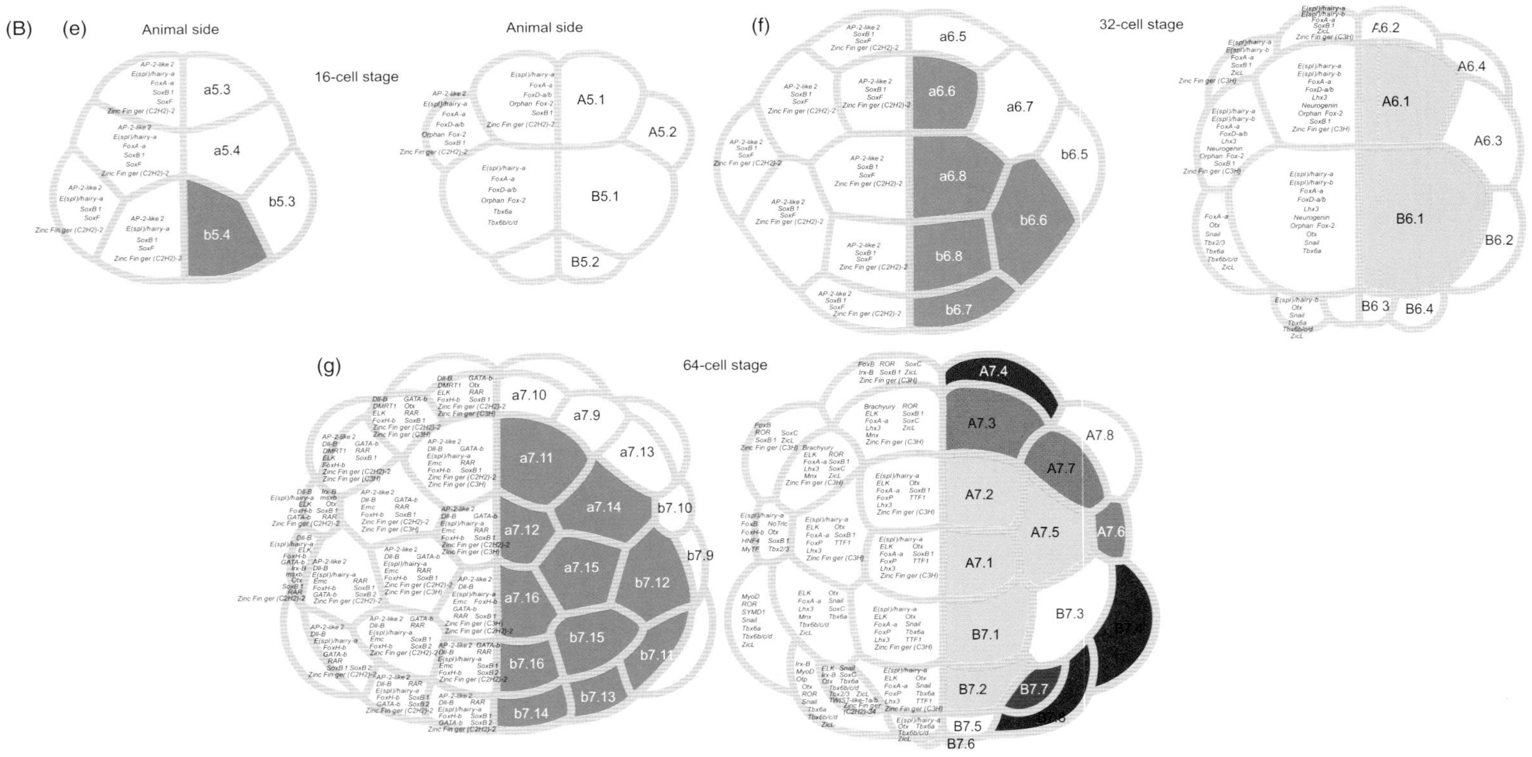

(B)

(e) Animal side

16-cell stage

Animal side

(f)

32-cell stage

64-cell stage

(g)

FIGURE 12.5 (Continued)

124

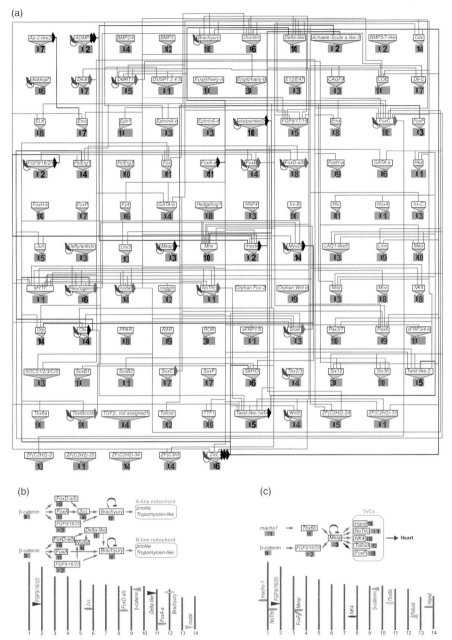

FIGURE 12.6   Gene regulatory networks responsible for the specification of *Ciona* embryonic cells, and the chromosomal localization of the genes. (a) The genes are ordered alphabetically from the top-left corner to the bottom-right corner. The chromosomal localization of genes is shown in light blue boxes, with short (left column) and long (right column) arms. Lines show gene networks, dark blue lines and arrows for same chromosome interactions, orange lines for inter-chromosomal interactions, and green arrows for auto regulatory interactions. Same chromosome interactions are highlighted. (b, c) The chromosomal localization of genes involved in the specification of (b) notochord cells and (c) adult heart cells. In the heart network (c), *Mesp* and *FoxF* are localized close together on the middle of the long arm of chromosome 3, but all the others are dispersed. (Adapted from Shoguchi *et al.* (2008) *Dev Biol*, **316**, 498–509.)

transcriptional network would contribute to our understanding of *Ciona* embryogenesis, and its evolutionary conservation between chordates.

Furthermore, systematic chromatin immunoprecipitation (ChIP) experiments for major nodes of the network have been carried out to disclose more comprehensive scheme of the networks. For example, 11 of the 15 *Ci-Bra* downstream genes that are specifically expressed in notochord cells have shown to be direct targets of Ci-Bra. These approaches may help identify conserved chordate linkages in regulatory networks. In parallel, networks act in a precise cellular context, and it will be important to integrate the transcriptional networks with the position and behavior of cells. This will certainly benefit from the small number of transparent cells in ascidian early embryos, whose precise individual morphology can be quantified in 3D reconstructed embryos shown in ANISEED.

The GRN in the *Ciona* embryo indicates a high prevalence of negative feedback loops and negative autoregulation. In total, 22 of the 27 genes examined negatively autoregulate themselves, directly or indirectly. For instance, *Brachyury* and *Nodal* is negatively autoregulated. In the case of *Nodal*, there is more than a 10-fold elevation in the levels of expression in embryos developed from eggs microinjected with a Nodal MO. This autoregulation may be critical to fine tune the amount of Nodal ligand presented to surrounding cells during comparatively prompt embryogenesis of ascidians.

As in the case of Nodal negative feedback, repression is an important feature of gene networks. For example, the Snail repressor is expressed in the notochord. Earlier, when *Brachyury* is first activated in the A-line at the 64-cell stage and B-line progenitors at the 112-cell stage, the Snail repressor is restricted to the trunk mesenchyme and developing tail muscles. *Snail* is activated in the tail muscles and helps exclude *Brachyury* expression from neighboring muscle cells when Delta-like and FGF9/16/20 induce notochord formation (Fig. 9.2b). However, by the onset of gastrulation, *Snail* is expressed in notochord cells, where it might attenuate *Brachyury* expression. Peak *Brachyury* expression in the notochord cells is seen at neurulation, but only low levels persist during elongation of the tail. It is conceivable that this downregulation is important for normal notochord differentiation, because sustained expression of high levels of *Brachyury* causes defects in notochord cell intercalation and tail elongation.

12.9 CHROMOSOMAL-LEVEL REGULATION OF GENE EXPRESSION

An intriguing question of embryology is how the temporal cascade of regulatory gene activation is controlled in relation to the division cycle of embryonic cells. In *Ciona* embryos, for instance, maternal β-catenin, and zygotic *FoxA, FGF9/16/20, EphrinAd, FoxD, ZicL, Nodal, Delta-like*, and *Brachyury* are regulatory genes involved in the network that forms the notochord. A key specifier *Brachyury* is expressed at the 64-cell stage in primordial notochord cells. It has been shown that the expression of *FoxD* initiates at the 16-cell stage as a direct target of β-catenin, *ZicL* is expressed at the 32-cell stage as a target of FoxD, and *Brachyury* is activated at the 64-cell stage as a target of ZicL. Namely, a relay of the gene expression is controlled at each division cycle of notochord precursor cells. Each division cycle of early *Ciona* embryos takes ~30 min. Judging from the time required for DNA synthesis for the next division and other factors associated with cell cycle, it is reasonable to assume that the transcription factors have to find out sequence-specific biding sites of their own targets within several minutes after the factor proteins is synthesized. Is gradual and gradient distribution of the factor protein synthesized in a given, restricted region of the nucleus sufficient to find and bind to target sites positioned in another region of the nucleus within this limited time period? Or is there any other device than simple physical gradient distribution of the factor proteins to make it easier to recognize the target site?

One possibility is a close positioning of a given TF gene and its target genes, which may explain why this short-time range and rigid gene activation is sufficient to find out the target site. However, as shown in Fig. 12.6b as to the *Brachyury* upstream cascade, all of these regulatory genes are found on different chromosomes: *FoxD* on the long arm of chromosome 8, *ZicL* on the long arm of chromosome 6, and *Brachyury* on the short arm of chromosome 12 (Fig. 12.6b). Therefore, there is no clustering of the regulatory genes dedicated to the formation of notochord, a most prominent feature of chordate embryos. Another example is the genetic cascade for differentiation of adult heart precursor cells, which includes maternal β-catenin and macho-1, and zygotic *Tbx6*, *FGF9/16/20*, *Mesp*, *Hand*, *NoTrlc*, *NK4*, *Tolloid*, and *FoxF* in this order (Fig. 12.6c). Although *Mesp* and *FoxF* are present on the long arm of chromosome 3, the other eight genes are distributed over seven different chromosomes, indicating a similar dispersion of the regulatory genes dedicated to the other major tissues of the early tadpole.

The circuit diagram for patterning of the early *Ciona* embryo from the 16-cell to the early gastrula stage contains 249-paired interactions between 27 upstream regulatory genes which are expressed at the 16- to 64-cell stages (including *NoTrlc*, *ADMP*, *FGF9/16/20*, *msxb*, and others; Fig. 12.7, left column) and 80 downstream regulatory genes which are expressed at the 32- to 110-cell stages (including *Eph1*, *SoxB1*, *COE*, *sFRP3/4-b*, *Hex*, and others; Fig. 12.7, right column). These 249-paired interactions include 14 on the same chromosome and 22 autoregulatory interactions, and the remaining 213 represent inter-chromosomal interactions (Fig. 12.7). The low ratio of the same chromosome interactions to nonautoregulatory interactions indicates upstream–downstream genes are dispersed on the chromosomes, while one of the upstream genes, *ZicL*, is involved in three same chromosome interactions (Fig. 12.7). There are several examples of fixed interactions from one chromosome to another. For example, an upstream gene (*FGF9/16/20*) on chromosome 2 regulates downstream genes on chromosomes 1 and 5, while *FoxA-a* on chromosome 11 regulates target genes on chromosome 1. However, in general, the interactions are distributed over the genome and no special tendency was found for chromosome-level interactions (Fig. 12.7).

Therefore, it is unlikely that the close chromosomal positioning of regulatory genes is required for genome-wide coordination of gene expression networks. However, a possibility still remains that the three dimensional architecture of chromatin of each of the 14-pair chromosomes, which may be modified at each division cycle, would allow close localization of the genes. This fundamental question of developmental biology should be examined in future in the *Ciona* system.

12.10   *Hox* GENES AND RETINOIC ACID

Hox genes of *C. intestinalis* lost their proper cluster organization (Fig. 12.8a). *Hox7, 8, 9,* and *11* are missing in the genome, and the gene cluster is split into two chromosomes (*Hox1–6* and *10* are on chromosome 1, and *Hox12* and *13* are on chromosome 7). *Hox10*

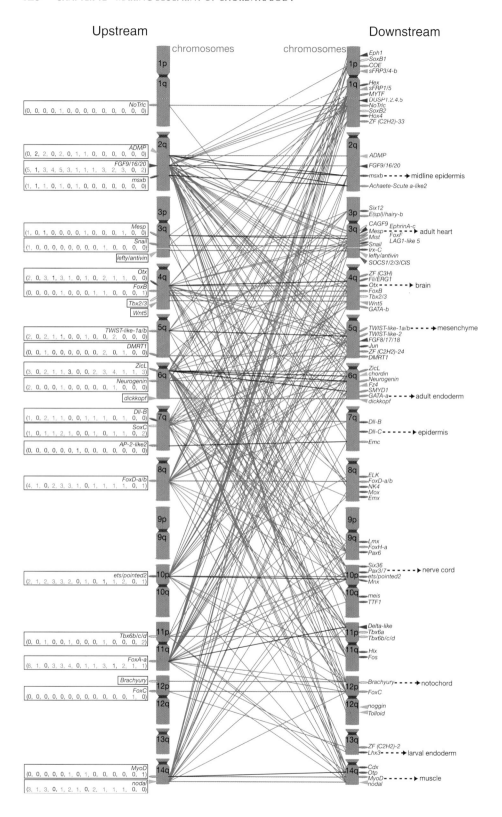

is located between *Hox2–4* and *Hox5–6*. However, even with the disrupted Hox gene cluster order, the collinear expression of these genes in the CNS is maintained in *C. intestinalis* (Fig. 12.8b, c). Accordingly, *Hox* genes have comparatively limited functions in the patterning of the *Ciona* CNS. *Hox10* is essential for CNS patterning and for generating the GABAergic/glycinergic neurons in the visceral (motor) ganglion, *Hox1* is required for the development of the atrial siphon placode, and *Hox12* is involved in the elongation of the larval tail tip.

The retinoic acid (RA)-mediated expression of *Hox1* is not only a hallmark of the CNS, but also functions in the epidermal ectoderm of chordates. RA is synthesized by *Raldh2* (retinaldehyde dehydrogenase 2), which is expressed in the most anterior muscle cells at the tailbud stage. *Ciona* mutant analyses show that the RA-*Hox1* system in the epidermal ectoderm is necessary for the formation of the atrial siphon placode, a structure homologous to the vertebrate otic placode. Loss of *Hox1* function or the abolishment of RA leads to a failure in atrial siphon placode formation, which can be rescued by expressing *Hox1* in the epidermis. Therefore, RA and *Hox1* in the epidermal ectoderm play key roles in the acquisition of the otic placode during chordate evolution. In addition, in the *Ciona* embryo, the antagonism of the RA and FGF/MAPK signals is required to control the antero-posterior patterning of the tail epidermis.

12.11   LEFT–RIGHT ASYMMETRY

The ascidian larvae do not swim in a straightforward manner, but rather rotate once every several tail beats. This characteristic movement is due to left–right (L–R) larval asymmetry. The asymmetry is distinguishable by the right positioning of the ocellus in the sensory vesicle. The ocellus pigment cell is first formed on the midline, and then moves toward the right side, followed by the formation of the photoreceptor cells on the right side.

In *Ciona* embryos, Nodal signaling is responsible for the L–R asymmetric formation of the ocellus. *Nodal* is expressed on the left side of the sensory vesicle at the late tailbud stage, and disruption of this expression results in symmetric formation of the ocellus. On the other hand, a paired class homeobox gene, *Rx* (*Ci-Rx*), is expressed bilaterally at the early tailbud stage, but the expression becomes repressed on the left side. At the late tailbud stage, *Ci-Rx* is expressed only on the right side of the sensory vesicle and regulates the expression of arrestin and opsin in the ocellus pigment cells and photoreceptor cells. Nodal signaling represses the expression of *Ci-Rx* on the left side of the embryo by the late tailbud stage, leading to subsequent formation of the photoreceptor cells on the right side while promoting a rightward shift of the ocellus pigment cell. Another unique

FIGURE 12.7   The genome-wide relations of regulatory genes involved in the network for the construction of the *Ciona* body plan. 249 chromosome-level relations between upstream regulatory genes (left column) and downstream regulatory genes (right) are indicated by lines. Chromosomes are shown as light blue boxes with a dark-blue centromere. The genes shown with dashed arrows (right side) are key genes for specification of major tissues of the *Ciona* embryo. The colors of discs and arrowheads are same as shown in Fig. 12.6. Number shown in parentheses indicates downstream gene number in order of 14 chromosomes. (Adapted from Shoguchi *et al.* (2008) *Dev Biol*, **316**, 498–509.)

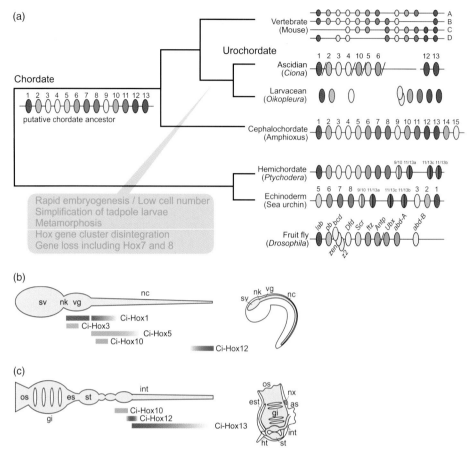

FIGURE 12.8 *Hox* gene clusters in *Ciona intestinalis*. (a) *Hox* gene organization in the deuterostome phylogeny. Colored ovals indicate *Hox* genes, and genes of smaller paralogous subgroup numbers are to the left and those of larger numbers to the right. Ovals on a line indicate a *Hox* gene cluster. Vertebrates have four *Hox* gene clusters, designated A through D, from the top to the bottom. *C. intestinalis Hox* genes have been mapped on two chromosomes. The putative gene order is shown for *Hox2-4* and *Ho5-6*. With respect to *Oikopleura Hox* genes, the chromosomal location is unknown. A putative chordate ancestor may possess a single *Hox* gene cluster with more or less than 13 member genes. Events likely acquired or occurred during the course of urochordate evolution are depicted in the sepia box. *Hox* gene organization in *Drosophila* is shown as an outgroup. (b, c) Expression domains of the *Hox* genes in *C. intestinalis*. The expression in the CNS during the tailbud stages (b) and in the juvenile gut (c) are shown schematically. The gut is drawn straightened to clearly indicate the anterior–posterior axis. Schematic drawings of the tailbud embryo and the juvenile are shown at the right. as, atrial siphon; es, esophagus; est, endostyle; gi, gill; ht, heart; int, intestine; nc, nerve cord; nk, neck; nx, neural complex; os, oral siphon; st, stomach; sv, sensory vesicle; vg, visceral ganglion. (Adapted from Ikuta and Saiga (2005) *Dev Dyn*, **233**, 382–389.)

feature of Nodal in *Ciona* embryos is that its L–R asymmetric expression is restricted to the ectoderm, not in the mesoendoderm as seen in vertebrate embryos.

The initial cue to establish L–R asymmetry in ascidian embryos is still unclear, but may involve the action of beating monocilia as in vertebrates. In a recent study of *Halocynthia* neurulae suggests that embryonic rotation is critical for the establishment of L–R asymmetry. Two hours prior to the onset of *Nodal* expression, the neurula embryo rotates along the anterior–posterior axis in a counterclockwise direction (when seen from a posterior view), and then this rotation stops when the left side of the embryo is oriented downwards. It is likely that epidermis monocilia, which appear at the neurula rotation stage, generate the driving force for this rotation. When the embryo lies on the left side, a protrusion of the neural fold physically prevents it from rotating further. Experiments in which neurula rotation is perturbed by various means indicate that contact of the left epidermis with the vitelline membrane as a consequence of neurula rotation promotes *Nodal* expression in the left epidermis. This suggests that chemical signals, rather than mechanical, from the vitelline membrane promote *Nodal* expression.

12.12 GENOMIC DNA METHYLATION AND CHROMATIN-LEVEL MODIFICATION OF REGULATORY GENES

During animal development, chromatin changes occur both globally and locally. At the global level, chromatin decondenses toward the characteristic "open" state of euchromatin. Local chromatin reorganization supports the activation and/or silencing of lineage-specific genes. The proteins that regulate this process act at different levels of chromatin regulation, including histone modifications, histone variants, chromatin remodeling, and genomic DNA methylation. As little is known on histone modifications and chromatin remodeling in ascidian embryos, genomic DNA methylation is discussed in relation to temporal control of gene expression.

An interesting suggestion on temporal control or clock mechanisms of the initiation of gene expression came from experiments carried out in 1980s by using ascidian embryos. That is, neither cytoplasmic division nor nuclear division of embryonic cells is required for development of tissue-specific differentiation markers but DNA replication is prerequisite for the marker development. When DNA replication of early ascidian embryos is pharmacologically blocked, the markers do not appear. In addition, a definite number of DNA replications are likely associated with the timing of muscle marker expression, suggesting that DNA replication cycle acts as clock mechanisms of gene expression. A suggested clock mechanism that counts the timing of the marker expression is DNA methylation. If demethylation of regulatory region of a given gene takes place every cell cycle (DNA replication), the grade of demethylation may be counted by the number of DNA replications, and after the DNA demethylation reaches a critical threshold level, sequence-specific binding of transcription factors to demethylated DNA may allow the gene expression.

In eukaryotes ranging from fungi, to plants to humans, DNA methylation is found exclusively at cytosine residues, and the overall pattern of DNA methylation varies in different organisms (Fig. 12.9A). In mammalian cells the pattern consists of long tracts of DNA in which CpGs are methylated to a high average level ($<80\%$) punctuated by short unmethylated regions called "CpG islands" (Fig. 12.9c). The most frequent

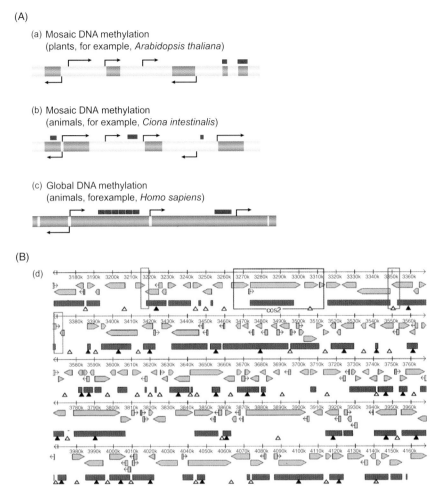

FIGURE 12.9 DNA methylation pattern in *Ciona intestinalis*. (A) The patterns in the plant *Arabidopsis thaliana* (a), *C. intestinalis* (b), and *Homo sapiens* (c). The upper two show different types of the mosaic methylation while the human being shows global methylation pattern with only CpG islands being unmethylated. (Modified from Suzuki and Bird (2008) *Nat Rev Genet*, **9**, 465–476). (B) (d) Methylated domains (MDs) within a 1-Mb genomic region from chromosome 4q of *Ciona intestinalis* genome colocalize with transcription units. MDs (green bars) are inferred from CpG[o/e] ratios. The inferred pattern is experimentally verified by methylation-sensitive PCR at sites marked by closed triangles (methylated) and open triangles (unmethylated). Yellow arrows show the size and orientation of predicted transcription units. The position of bisulfite sequenced cosmid insert cos2 is shown by black boxes. Green boxes indicate regions that are bisulfite sequenced in order to test the accuracy of predicted unmethylated domain (UMD)/MD boundaries. (e) Frequency of predicted genes that overlap to varying degrees with MDs (left). (Middle) Frequency of MDs that overlap to varying degrees with predicted genes. (Right) Frequency of UMDs with gene overlap. (Modified from Suzuki *et al*. (2007) *Genome Res*, **17**, 625–631.)

(e)

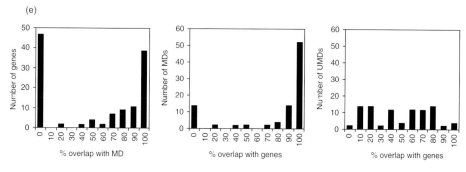

FIGURE 12.9   (*Continued*)

pattern in invertebrates is "mosaic methylation," comprising domains of heavily methylated DNA interspersed with domains that are methylation free (Fig. 12.9b), while the plant *Arabidopsis thaliana* display a mosaic DNA methylation pattern that is reminiscent of invertebrate animals (Fig. 12.9a).

The distribution of DNA methylation in the *C. intestinalis* genome has been examined by bisulfite sequencing and computational analysis. It has been shown that methylated domains with sharp boundaries strongly colocalize with <60% of transcription units (Fig. 12.9B). By contrast, promoters, intergenic DNA, and transposons are not preferentially targeted by DNA methylation. Methylated transcription units include evolutionarily conserved genes, whereas the most highly expressed genes preferentially belong to the unmethylated fraction. These results lend support to the hypothesis that CpG methylation functions to suppress spurious transcriptional initiation within infrequently transcribed genes. The pattern of DNA methylation is constant in all constituent cells of *C. intestinalis* and does not change during the embryogenesis. Therefore, the likelihood that DNA methylation functions as a clock mechanism is quite low. The meaning of this "body pattern-like" DNA methylation of the *Ciona* genome remains to be elucidated.

12.13   MICRO RNAs

Micro RNAs (miRNAs) are small ~21-nucleotide noncoding RNAs implicated in diverse biological processes in a wide range of metazoans. miRNAs bind to mRNA transcripts and downregulate their expression either through mRNA destabilization or translational repression. Recent computational studies have identified a total of 375 or more known miRNAs in the *C. intestinalis* genome.

Of them, miR-124 is one of the most abundant miRNAs that are expressed in the nervous system of mice, *Drosophila* and *C. elegans*, and plays a role in regulation of two antineural transcription factors. Similarly, *C. intestinalis* miR-124 plays a role in promoting nervous system development by feedback interaction between miR-124 and Notch signaling that regulates the epidermal-peripheral nervous system (PNS) fate choice in tail midline cells. Notch signaling silences *miR-124* in epidermal midline cells, whereas in PNS midline cells miR-124 silences *Notch, Neuralized*, and *Hairy/Enhancer-of-Split* genes (Fig. 12.10a).

In addition, more broadly, genome-wide target extraction with validation using an in vivo tissue-specific sensor assay indicates that miR-124 shapes neuronal progenitor fields by downregulating nonneural genes, notably the muscle specifier Macho-1

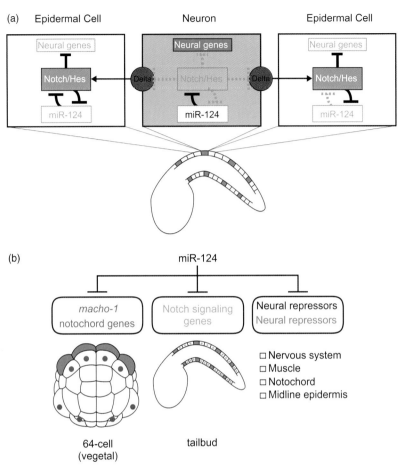

FIGURE 12.10   Model of miR-124 role in *Ciona* neuronal development. (a) Model for the interaction between miR-124 and Notch signaling along the tail midlines. In epithelium-sensory neurons (ESNs), *miR-124* expression is predicted to silence Notch signaling, whereas in non-ESN cells Notch activation results in the repression of *miR-124* expression. If Notch signaling is blocked, ectopic ESNs expressing ectopic *miR-124* are formed along the midlines. (b) Model for miR-124 function based on the expression pattern of *miR-124* (cells shaded green) and the characterization of identified miR-124 target genes in *Ciona*. Relevant tissues and associated pathways are color-coded. Note that nervous system expression is more extensive than shown (only neuronal cells relevant to targeted pathways are shaded green). Notochord genes refer to those genes that are modulated by the notochord specifier *Brachyury*. The right-hand list includes targeted genes and pathways that cannot confidently be assigned to a particular embryo stage. (Adapted from Chen *et al.* (2011) *Development* **138**, 4943–4953.)

and fifty Brachyury-regulated notochord genes, as well as several antineural factors including SCP1 and PTBP1 (Fig. 12.10b). The 3′UTR conservation analysis reveals that miR-124 targeting of SCP1 is likely to have arisen as a shared, derived trait in the vertebrate/tunicate ancestor and targeting of PTBP1 is conserved among bilaterians except for ecdysozoans, while extensive Notch pathway targeting appears to be *Ciona* specific. Taking advantage of *Ciona* system, future studies of ascidian miRNAs may facilitate our understanding of the expression and function of miRNA genes.

SELECTED REFERENCES

BERTRAND, V., HUDSON, C., CAILLOL, D., POPOVICI, C. *et al.* (2003) Neural tissue in ascidian embryos is induced by FGF9/16/20, acting via a combination of maternal GATA and Ets transcription factors. *Cell*, **115**, 615–627.

CHEN, J.S., SAN PEDRO, M. ZELLER, R.W. (2011) miR-124 function during Ciona intestinalis neuronal development includes extensive interaction with the Notch signaling pathway. *Development*, **138**, 4943–4953.

CHRISTIAEN, L., STOLFI, A., DAVIDSON, B., LEVINE, M. (2009) Spatio-temporal intersection of Lhx3 and Tbx6 defines the cardiac field through synergistic activation of Mesp. *Dev Biol*, **328**, 552–560.

DARRAS, S., NISHIDA, H. (2001) The BMP signaling pathway is required together with the FGF pathway for notochord induction in the ascidian embryo. *Development*, **128**, 2629–2638.

DAVIDSON, B., SHI, W., BEH, J., CHRISTIAEN, L. *et al.* (2006) FGF signaling delineates the cardiac progenitor field in the simple chordate, *Ciona intestinalis*. *Genes Dev*, **20**, 2728–2738.

HUDSON, C., DARRAS, S., CAILLOL, D. (2003) A conserved role for the MEK signalling pathway in neural tissue specification and posteriorisation in the invertebrate chordate, the ascidian *Ciona intestinalis*. *Development*, **130**, 147–159.

HUDSON, C., LOTITO, S., YASUO, H. (2007) Sequential and combinatorial inputs from Nodal, Delta2/Notch and FGF/MEK/ERK signalling pathways establish a grid-like organisation of distinct cell identities in the ascidian neural plate. *Development*, **134**, 3527–3537.

HUDSON, C., YASUO, H. (2006) A signalling relay involving Nodal and Delta ligands acts during secondary notochord induction in *Ciona* embryos. *Development*, **133**, 2855–2864.

HUDSON, C., YASUO, H. (2005) Patterning across the ascidian neural plate by lateral Nodal signalling sources. *Development*, **132**, 1199–1210.

IKUTA, T., YOSHIDA, N., SATOH, N., SAIGA, H. (2004). *Ciona intestinalis* Hox gene cluster: its dispersed structure and residual colinear expression in development. *Proc Natl Acad Sci USA*, **101**, 15118–15123.

IKUTA, T., SATOH, N., SAIGA, H. (2010) Limited functions of Hox genes in the larval development of the ascidian *Ciona intestinalis*. *Development*, **137**, 1505–1513.

IMAI, K.S., HINO, K., YAGI, K., SATOH, N. *et al.* (2004) Gene expression profiles of transcription factors and signaling molecules in the ascidian embryo: towards a comprehensive understanding of gene networks. *Development*, **131**, 4047–4058.

IMAI, K.S., LEVINE, M., SATOH, N., SATOU, Y. (2006) Regulatory blueprint for a chordate embryo. *Science*, **312**, 1183–1187.

IMAI, K.S., STOLFI, A., LEVINE, M., SATOU, Y. (2009) Gene regulatory networks underlying the compartmentalization of the *Ciona* central nervous system. *Development*, **136**, 285–293.

KIM, G.J., YAMADA, A., NISHIDA, H. (2000) An FGF signal from endoderm and localized factors in the posterior–vegetal egg cytoplasm pattern the mesodermal tissues in the ascidian embryo. *Development*, **127**, 2853–2862.

KIM, G.J., KUMANO, G., NISHIDA, H. (2007) Cell fate polarization in ascidian mesenchyme/muscle precursors by directed FGF signaling and role for an additional ectodermal FGF antagonizing signal in notochord/nerve cord precursors. *Development*, **134**, 1509–1518.

KONDOH, K., KOBAYASHI, K., NISHIDA, H. (2003) Suppression of macho-1-directed muscle fate by FGF and BMP is required for formation of posterior endoderm in ascidian embryos. *Development*, **130**, 3205–3216.

KUMANO, G., YAMAGUCHI, S., NISHIDA, H. (2006) Overlapping expression of FoxA and Zic confers responsiveness to FGF signaling to specify notochord in ascidian embryos. *Dev Biol*, **300**, 770–784.

LEMAIRE, P. (2009) Unfolding a chordate developmental program, one cell at a time: invariant cell lineages, short-range inductions and evolutionary plasticity in ascidians. *Dev Biol*, **332**, 48–60.

LEMAIRE, P., SMITH, W.C., NISHIDA, H. (2008) Ascidians and the plasticity of the chordate developmental program. *Curr Biol*, **18**, R620–R631.

MINOKAWA, T., YAGI, K., MAKABE, K.W., NISHIDA, H. (2001) Binary specification of nerve cord and notochord cell fates in ascidian embryos. *Development*, **128**, 2007–2017.

NAKAMURA, Y., MAKABE, K.W., NISHIDA, H., 2005. POPK-1/Sad-1 kinase is required for the proper translocation of maternal mRNAs and putative germ plasm at the posterior pole of the ascidian embryo. *Development*, **132**, 4731–4742.

NISHIDE, K., MUGITANI, M., KUMANO, G., NISHIDA, H. (2012) Neurula rotation determines left-right asymmetry in ascidian tadpole larvae. *Development*, **139**, 1467–1475.

PASINI, A., MANENTI, R., ROTHBACHER, U., LEMAIRE, P. (2012) Antagonizing retinoic acid and FGF/MAPK pathways control posterior body patterning in the invertebrate chordate *Ciona intestinalis*. *PLoS One*, **7**, e46193.

PICCO, V., HUDSON, C., YASUO, H. (2007) Ephrin−Eph signalling drives the asymmetric division of noto-chord/neural precursors in *Ciona* embryos. *Development*, **134**, 1491–1497.

SASAKURA, Y., KANDA, M., IKEDA, T., HORIE, T. *et al.* (2012) Retinoic acid-driven Hox1 is required in the epidermis for forming the otic/atrial placodes during ascidian metamorphosis. *Development*, **139**, 2156–2160.

SATOU, Y., SATOH, N., IMAI, K.S. (2008) Gene regulatory networks in the early ascidian embryo. *Biochim Biophys Acta*, **1789**, 268–273.

SHI, W., LEVINE, M. (2008) Ephrin signaling establishes asymmetric cell fates in an endomesoderm lineage of the *Ciona* embryo. *Development*, **135**, 931–940.

SHOGUCHI, E., HAMAGUCHI, M., SATOH, N. (2008) Genome-wide network of regulatory genes for construction of a chordate embryo. *Dev Biol*, **316**, 498–509.

SUZUKI, M.M., KERR, A.R.W., SOUSA, D.D., BIRD, A. (2007) CpG methylation is targeted to transcription units in an invertebrate genome. *Genome Res*, **17**, 625–631.

SUZUKI, M.M., BIRD, A. (2008) DNA methylation landscapes: provocative insights from epigenomics. *Nat Rev Genet*, **9**, 465–476.

TASSY, O., DAIAN, F., HUDSON, C., BERTRAND, V. *et al.* (2006) A quantitative approach to the study of cell shapes and interactions during early chordate embryogenesis. *Curr Biol*, **16**, 345–358.

TASSY, O., DAUGA, D., DAIAN F., SOBRAL, D. *et al.* (2010) The ANISEED database: digital representation, formalization, and elucidation of a chordate developmental program. *Genome Res*, **20**, 1459–1468.

YAGI, K., SATOH, N., SATOU, Y. (2004) Identification of downstream genes of the ascidian muscle determinant gene Ci-macho1. *Dev Biol*, **274**, 478–489.

YOSHIDA, K., SAIGA, H. (2011). Repression of Rx gene on the left side of the sensory vesicle by Nodal signaling is crucial for right-sided formation of the ocellus photoreceptor in the development of *Ciona intestinalis*. *Dev Biol*, **354**, 144–150.

CHAPTER 13

DEVELOPMENT OF THE JUVENILE HEART

Ascidians develop three types of muscles: larval tail muscle (Chapter 6), adult body-wall (siphon and mantle) muscle, and adult heart muscle. The larval tail muscle cells are unicellular and striated. The body-wall muscle cells are multinuclear and smooth. The heart muscle cells are unicellular and striated. As the discovery of *tinman* in *Drosophila* and its vertebrate homolog *Nkx2.5*, the genetic regulatory network involved in heart development has been extensively documented using both flies and vertebrates. A circuit consisting of *GATA*, *Nkx*, and *Hand* is highly evolutionarily conserved and constitutes a so-called kernel. *Ciona* heart development provides another attractive experimental system to investigate the molecular and cellular mechanisms from the beginning of fertilization to the final stage of heart formation at the single cell level.

13.1 THE ASCIDIAN HEART

Ascidians have an open circulatory system. Their short, tubular heart is located postventrally in the adult body near the stomach and behind the pharyngeal basket (Fig. 1.1m, o and Fig. 13.1a). The heart consists of a valveless myocardial tube encased within a pericardial coelom (Fig. 13.1b). In *Ciona* and many other ascidians, the myocardial tube is V shaped, fitting into a roughly triangular pericardial sac (Fig. 13.1b). The pericardium and myocardium are derived from a single continuous epithelial tube invaginated along the dorsal side. Excitation–contraction coupling appears to rely on peripheral couplings, or diads, as in vertebrate cardiac muscle. The junctions between the myoepithelial cells bear some resemblance to the intercalated discs of the vertebrate myocardium. There is no discernible endocardium in the *Ciona* heart. Instead, the lumen of the heart is lined with the basal lamina of the myoepithelial cells. Along the junction of the pericardial and myocardial tubes there is a dense extracellular matrix, termed the raphe (Fig. 13.1b). Opposing the raphe is a line of smaller noncontractile cells, termed the undifferentiated line (Fig. 13.1b).

The ascidian heart drives blood circulation through peristalsis. The striated myofibrils along the wall of the heart run parallel to each other and at a $60–70°$ angle to the long axis of the heart. Thus, these fibers form a "tight spiral around the heart" and rhythmic contractions proceed in a gradual wringing motion. This may be critical to preventing backflow, in a manner similar to the recently described spiral wringing of the early vertebrate heart tube. Regular contractile waves can originate at either end of the heart,

Developmental Genomics of Ascidians, First Edition. Noriyuki Satoh.
© 2014 John Wiley & Sons, Inc. Published 2014 by John Wiley & Sons, Inc.

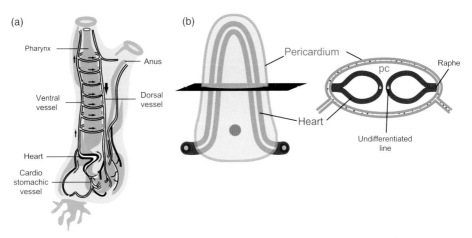

FIGURE 13.1   Diagram of the adult *Ciona* heart. (a) The position of the heart and the circulatory flow (red and black arrows) in the adult body. (b) Cartoon (left) and cross-sectional diagram (right) of the *Ciona* heart. The myocardial tube (red), pericardium (blue), pericardial coelom (pc), undifferentiated line, and raphe are indicated. (Modified from Davidson (2007) *Semin Cell Dev Biol*, **18**, 16–26.)

driving periodic reversals in blood flow. The core mechanism for this reversibility is the presence of two myogenic pacemakers, each one at both the ends of the heart. Alternating predominance of one or the other appears to determine the direction of peristalsis.

13.2   LINEAGE OF CELLS INVOLVED IN HEART DEVELOPMENT

A pair of B7.5 cells, with a documented lineage, in the 112-cell embryo is the starting point for heart development (Fig. 13.2a). B7.5 cells specifically express the bHLH transcription factor gene, *Mesp*. The mouse has two *Mesp* genes that display overlapping expression and have redundant functions during muscle development, making research into their individual roles difficult. By contrast, *Ciona* bears only a single copy of the gene. Larvae that have reduced *Mesp* expression metamorphose into outwardly normal looking juveniles that completely lack a heart, although the heartless, *tinman*-like juveniles can survive for a few weeks.

The B7.5 lineage proceeds in the following manner. B7.5 first divides equally into B8.9 and B8.10 in the early gastrula (Fig. 13.2b). Then, in the late gastrula/neurula, B8.9 and B8.10 divide into smaller anterior (rostral) daughter cells and larger posterior (caudal) daughter cells, respectively (Fig. 13.2c). During larval tail extension, the rostral cells separate from their caudal sisters and become TVC, while the caudal cells form anterior muscle cells (ATMs) of the larval tail (Fig. 13.2d). The ATMs remain in their initial position and undergo no further cell divisions or migration. By contrast, the TVCs migrate anteriorly along the ventral surface of the trunk endoderm by extending filopodia that meets along the ventral midline (Fig. 13.2e). Thus, TVCs exhibit two phases of directed cell migration—anterior movement along the ventral endoderm followed by midline positioning. After meeting along the midline, the TVCs initiate asymmetric cell divisions. The first round of division leads to a bilinear cluster of eight cells, four large outer cells and four small inner cells. After hatching, the four outer cells undergo another

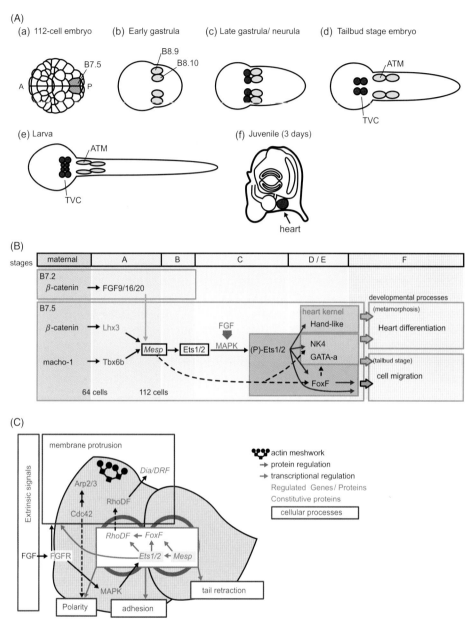

FIGURE 13.2 Development of the *Ciona* heart including the genetic regulatory cascade. (A) (a–f) Proliferation and specification of B7.5 at the 110-cell stage. The cell gives rises to trunk ventral cells (TVC, red) and larval anterior tail muscle cells (ATM, yellow). TVCs form juvenile heart (red) after metamorphosis. A, anterior; P, posterior. (Modified from Satou *et al.* (2004) *Development*, **131**, 2533–2541.) (B) The gene regulatory network (GRN) underlying *Ciona* heart formation. Almost all key players involved in the process are described. (Modified from Christiaen *et al.* (2009) *Dev Biol*, **328**, 552–560 and Beh *et al.* (2007), *Development*, **134**, 3297–3305.) (C) The GRN influences polarity, adhesion, membrane protrusion, and tail retraction of constituent cells. Membrane protrusion is controlled in part by *FoxF*- and *Ets1/2*-mediated upregulation of *RhoDF*, which functions together with constitutively expressed *Cdc42*, *Dia/DRF* (red square), and *Arp2/3* (red square). See the text for details. (Adapted from Christiaen *et al.* (2008) *Science*, **320**, 1349–1352.)

round of asymmetric division leading to a total of 12 cells, four larger outer cells and eight smaller inner cells. Together, they form a flat sheet. In early-stage juveniles, the heart rudiment consists of a seemingly unstructured mass of cells, from which the heart tube later develops.

13.3   THE GENE REGULATORY NETWORK

Taking full advantage of the *Ciona* experimental model system presented in Chapters 2–4, the juvenile heart muscle development-related GRN, including factors upstream and downstream of *Mesp*, has been extensively profiled and is discussed in detail in the following sections.

13.3.1   The Upstream Cascade

Two maternally provided factors and two zygotically expressed genes are cell-autonomously responsible for *Mesp* expression in B7.5 cells. The minimal cardiac enhancer of *Mesp* features a T-box transcription factor Tbx6 binding site for the direct activation of gene expression. As *Tbx6* is downstream of maternal *macho-1*, the *macho-1/Tbx6/Mesp* triad represents an upstream signaling pathway (Fig. 13.2B). However, as *Tbx6* is expressed throughout the presumptive tail muscle cells, it alone cannot account for the restricted expression of *Mesp* in B7.5. Another possible activator is the LIM-homeobox gene, *Lhx3* that is a direct target of maternal β-catenin and is expressed throughout the presumptive endoderm and in B7.5 (Fig. 13.2B). Notably, B7.5 is the only cell to express sustained levels of both *Tbx6* and *Lhx3*. As misexpression of *Lhx3* is sufficient to induce ectopic *Mesp* activation in cells expressing *Tbx6*, it is likely that the two factors work synergistically to activate *Mesp* in B7.5. In addition, a noncell-autonomous FGF signaling pathway is also involved in *Mesp* activation. FGF9/16/20 is produced by the neighboring B7.2 cell (primordial endoderm). *FGF9/16/20* is a target of maternal β-catenin and functions together with *Tbx6* and *Lhx3* to activate *Mesp* (Fig. 13.2B).

13.3.2   The Downstream Cascade

The expression of *Ets1/2*, a transcriptional effector of RTK signaling, initiates during gastrulation in B8.9 and B8.10, daughter cells of the *Mesp*-expressing B7.5 (Fig. 13.2B). *Ets1/2* is likely to act downstream of *Mesp*. After gastrulation, B8.9 and B8.10 divide asymmetrically into TVCs and ATMs, respectively. TVCs activate RTKs by phosphorylating ERK, while the ATMs do not (Fig. 13.2B). General inhibition of RTK signaling, or the targeted inhibition of *Ets1/2* function blocks heart specification. Conversely, the targeted expression of a constitutively active form of *Ets1/2* induces ATMs to form heart cells (discussed below). The asymmetry of RTK activation in the rostral and caudal daughter cells is induced by FGF signaling through MAPK (Fig. 13.2B). *FGF9/16/20* is expressed in cells adjacent to the rostral daughters. Targeted inhibition of FGF receptor function blocks heart specification. In addition, TVCs employ polarized, invasive protrusions to localize their response to an ungraded signal. In association with enhanced protrusive activity toward underlying epidermal cells, TVCs gain enriched receptor activation along the ventral membrane, which eventually lead differential expression of heart progenitor genes.

Another gene that plays a significant role in *Ciona* heart development is *FoxF* (Fig. 13.2B). *FoxF* is one of the genes directly activated by Ets1/2 (phosphorylated ERK) in TVCs in response to FGF signaling. Interestingly, the targeted expression of a dominant-negative form of *FoxF* inhibits cell migration but not heart differentiation, resulting in a striking phenotype—a beating heart at an ectopic location within the juvenile body cavity. Accordingly, TVC migration requires *FoxF* function downstream and in parallel to the FGF/MAPK/Ets cascade (Fig. 13.2B). A combination of techniques, including cell sorting, microarray analysis, and targeted molecular manipulation were used to gain additional insights into mechanisms underlying cardiac muscle cell migration. These studies determined that FGF and FoxF directly upregulate the small guanosine triphosphatase RhoDF, which synergizes with Cdc42 to contribute to the protrusive activity of migrating cells (Fig. 13.2C). Moreover, RhoDF induces membrane protrusions independently of other cellular activities that are required for migration (Fig. 13.2C).

In parallel with the FGF/MAPK/Ets/FoxF cascade, Mesp regulates *GATA, Nkx (NK4), HAND*, and *NoTrlc*, the heart regulatory kernel in fly and vertebrates (Fig. 13.2B). This represents another instance where the *Ciona* model system provides more precise information with respect to gene functions during heart development. A coordination of cell migration and gene expression plays a pivotal role in precardiac muscle cell differentiation. Namely, migrating TVCs experience increasing BMP signaling as they migrate toward the ventral trunk epidermis that expresses sustained levels of *Bmp2/4*. This increasing signaling intensity allows the successive activation of *GATAa, Tolloid, Bmp2/4*, and *NK4*. Initial activation of *GATAa, Tolloid*, and *Bmp2/4* contributes to a positive feedback loop involving cell migration, chordin inhibition, and BMP ligand production. Sustained levels of BMP signaling become sufficient to activate *NK4* expression, which in turn contributes to a negative feedback loop inhibiting *Bmp2/4* and *Tolloid* expression. In addition, NK4 appears to inhibit cell migration thus providing a "transcriptional brake" to stop TVC migration.

GATA family transcription factors are core components of the vertebrate heart gene network. Although vertebrate GATA factors contribute to heart formation indirectly, by regulating endoderm morphogenesis, the precise impact of GATA factors on vertebrate cardiogenesis is masked by functional redundancies within multiple lineages. *Ciona GATAa* is expressed in the heart progenitor cells and adjacent endoderm. Targeted repression of *GATAa* activity in the heart progenitors changes their transcriptional identity. Targeted repression of endodermal *GATAa* function disrupts endoderm morphogenesis, such that the bilateral heart progenitors fail to fuse at the ventral midline. The resulting phenotype is strikingly similar to cardia bifida, a defect observed in vertebrate embryos, which results when endoderm morphogenesis is disturbed. Therefore, *Ciona GATAa* recapitulates cell-autonomous and noncell-autonomous roles performed by multiple, redundant GATA factors in vertebrate cardiogenesis.

13.4   CHANGES IN CHORDATE HEART MORPHOLOGY THROUGH EVOLUTION

As mentioned above, the ascidian heart is a simple tube-like organ, while the vertebrate heart has chambers, partitioned by septa, which sequentially contract under the control of the cardiac conduction system. It is highly likely that vertebrates evolved the dual-chambered heart through modification of the gene regulatory networks responsible for heart formation in ancestral chordates.

A provocative hypothesis with respect to vertebrate dual-chambered heart development comes from an experiment in which targeted expression of constitutively activated *Ets1/2* converted larval tail muscle progenitors to a heart fate, resulting in a doubling of the heart field. Some of the resulting juveniles had a heart with two compartments that beat sometimes independently and sometimes coordinately. This two-compartment heart lacked some features that characterize vertebrate hearts, that is, a conduction system, valves, and septa.

Although this two-compartment mutant heart needs to be further characterized morphologically, anatomically, developmentally, and electrophysiologically, it has engendered a novel hypothesis that could alter how we think about the evolution of the chambered heart. Specifically, an increase in the number of primordial heart cells can be obtained by recruiting competent cells that normally follow an alternative fate, rather than by excess cell divisions. Excess cell division does not expand the net volume of cells following a given fate, whereas cell recruitment does. Cell recruitment through cell fate transformation could add new components to an existing structure without impairing its function. In addition, such recruitment could be obtained though subtle changes in a genetic program, such as altered expression of *FGF9/16/20*. It is therefore possible that this strategy represents a general pathway of evolutionary change. It is not difficult to imagine that the vertebrate multi-chambered heart evolved by the recruitment of competent cells that occurred after a subtle change in FGF signaling, although many issues will need to be addressed in order to confirm this scenario.

Another evolutionary innovation of the vertebrate heart is the presence of two heart fields (Fig. 13.3). The vertebrate heart initially forms as a tube from a population of

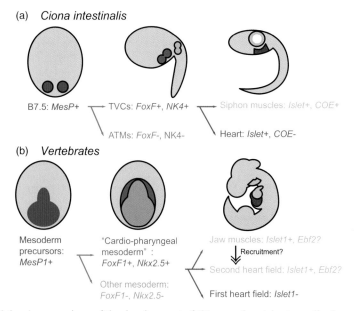

FIGURE 13.3   A comparison of the development of *Ciona* and vertebrate cardiopharyngeal mesoderm. (a) Summary of differential expression of selected regulatory genes in the *Ciona* B7.5 lineage. (b) Expression of orthologous genes in mouse heart development. Evolutionary reallocation of Islet1+ cardiopharyngeal precursors toward the heart might have given rise to the secondary heart field in vertebrates. (Adapted from Stolfi *et al.* (2010) *Science*, **329**, 565–568.)

precursor cells termed the first heart field (FHF). Cells from the adjacent, second heart field (SHF) are then progressively added to the developing heart. In avian and mammalian hearts, the FHF contributes mainly to the left ventricle, whereas the SHF gives rise to the outflow tract and large portions of the right ventricle and atria. Both fields arise from common mesodermal progenitors, although the detailed lineage relationships between the FHF and SHF remain uncertain. SHF-like territories have been identified in frog, zebrafish, and lamprey, yet evidence for a deeper evolutionary origin remains obscured by the absence of a clear SHF in invertebrates.

The *Ciona* heart progenitor cells also generate precursors of the atrial siphon muscles (ASMs) (Fig. 13.3a). These precursors express *Islet* and *Tbx1/10*, evocative of the splanchnic mesoderm that produces the lower jaw muscles and SHF of vertebrates (Fig. 13.3b). In addition, the transcription factor, *COE*, is a critical determinant of ASM fate (Fig. 13.3a). Therefore, it is likely that the last common ancestor of tunicates and vertebrates possessed multipotent cardiopharyngeal muscle precursors, and their reallocation might have contributed to the emergence of the SHF.

In summary, studies of the relatively simple *Ciona* gene regulatory network in the TVC, which depicts a system that was established prior to the gene duplication seen in vertebrate lineages, provide critical insights into the related but much more complex vertebrate heart progenitor networks.

SELECTED REFERENCES

BEH, J., SHI, W., LEVINE, M., DAVIDSON, B. *et al.* (2007) *FoxF* is essential for FGF-induced migration of heart progenitor cells in the ascidian *Ciona intestinalis*. *Development*, **134**, 3297–3305.

CHRISTIAEN, L., DAVIDSON, B., KAWASHIMA, T., POWELL, W. *et al.* (2008) The transcription/migration interface in heart precursors of *Ciona intestinalis*. *Science*, **320**, 1349–1352.

CHRISTIAEN, L., STOLFI, A., DAVIDSON, B., LEVINE, M. (2009) Spatio-temporal intersection of *Lhx3* and *Tbx6* defines the cardiac field through synergistic activation of *Mesp*. *Dev Biol*, **328**, 552–560.

CHRISTIAEN, L., STOLFI, A., LEVINE, M. (2010) BMP signaling coordinates gene expression and cell migration during precardiac mesoderm development. *Dev Biol*, **340**, 179–187.

COOLEY, J., WHITAKER, S., SWEENEY, S., FRASER, S. *et al.* (2011) Cytoskeletal polarity mediates localized induction of the heart progenitor lineage. *Nat Cell Biol*, **13**, 952–957.

DAVIDSON, B. (2007) *Ciona intestinalis* as a model for cardiac development. *Semin Cell Dev Biol* **18**, 16–26.

DAVIDSON, B., SHI, W., BEH, J., CHRISTIAEN, L. *et al.* (2006) FGF signaling delineates the cardiac progenitor field in the simple chordate, *Ciona intestinalis*. *Genes Dev*, **20**, 2728–2738.

SATOU, Y., IMAI, K.S., SATOH, N.. (2004) The ascidian *Mesp* gene specifies heart precursor cells. *Development* **131**, 2533–2541.

SIMOES-COSTAA, M.S., VASCONCELOSA, M., SAMPAIOA, A.C., CRAVO, R.M. *et al.* (2005) The evolutionary origin of cardiac chambers. *Dev Biol*, **277**, 1–15.

SRIVASTAVA, D. (2006) Making or breaking the heart: from lineage determination to morphogenesis. *Cell*, **126**, 1037–1048.

STOLFI, A., GAINOUS, B., YOUNG, J.J., MORI, A. *et al.* (2010) Early chordate origins of the vertebrate second heart field. *Science*, **329**, 565–568.

ZAFFRAN, S., KELLY, R.G. (2012) New developments in the second heart field. *Differentiation*, **84**, 17–24.

CHAPTER 14

GERM-CELL LINE, GAMETES, FERTILIZATION, AND METAMORPHOSIS

This chapter discusses recent advances in our understanding of germ cell formation, gametogenesis, fertilization, and metamorphosis of ascidians. *Ciona intestinalis* provides an attractive experimental system to study cellular and molecular mechanisms involved in the above processes.

14.1 GERM CELL FORMATION IN *Ciona*

In metazoans, genetic information is transmitted to subsequent generations only through germ cells. Owing to the evolutionary significance of the germline, many animals, including *Drosophila* and *Xenopus*, have adopted a developmental strategy whereby the germline is separated from the somatic line very early in embryogenesis. The germline fate is initially encoded in a small group of cells, the presumptive primordial germ cells (PGCs). Presumptive PGC fate is specified by the inheritance of the so-called germplasm, which is a distinctive granular egg cytoplasm containing germline determinants located at one pole of the zygote. In contrast to this sequestering strategy, mammals and urodele amphibians use inductive signaling to form presumptive PGCs. *Ciona* provides an experimental system to study both mechanisms in a single species.

14.1.1 Primordial Germ Cell Specification

In *C. intestinalis*, cells exhibiting the ultrastructural characteristics of PGCs appear in the developing juvenile gonad at around 6–10 days after metamorphosis (Chapter 2). The *Vasa* gene that encodes an RNA-binding protein shows germline-specific expression in a wide range of animals. After fertilization, the *Ciona Vasa* maternal mRNA and protein are both highly concentrated in the posterior-vegetal blastomeres, and finally localize to B7.6 of the 112-cell embryo (Fig. 14.1a, b, n). After metamorphosis, these *Vasa*-positive cells are incorporated into the primitive juvenile gonad to form germ cells (Fig. 14.1g–m), suggesting that the posterior-vegetal embryonic cells are presumptive PGCs. This is also the case in *Halocynthia roretzi* and *Phallusia mammillata*.

Developmental Genomics of Ascidians, First Edition. Noriyuki Satoh.
© 2014 John Wiley & Sons, Inc. Published 2014 by John Wiley & Sons, Inc.

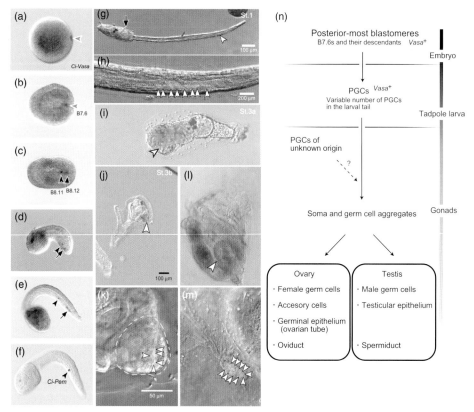

FIGURE 14.1 The expression profile of *Ci-Vasa* during *Ciona* embryogenesis. (a–e) *Ci-Vasa* RNA is found in (a) the posterior region of the 1-cell embryo, (b) B7.6 of the gastrula, (c) B8.11 and B8.12 in the neurula, (d) the early tailbud, and (e) the late tailbud embryo. (f) *Ci-pem* RNA expression in comparison to *Ci-Vasa*. Arrowheads in c–e indicate B8.11-line cells and arrows B8.12 line. (g, h) An early tadpole larva. Eight *Ci-Vasa*-positive cells are localized to the distal tail (arrowheads in h). *Ci-Vasa* signals are also detected in the trunk (arrow). (i) A stage 3a juvenile. *Ci-Vasa*-positive cells are among those in the tail debris (arrowhead). (j, k) A stage 3b juvenile. *Ci-Vasa*-positive cells (arrowheads) remain among those in the tail debris (outlined). (l, m) A stage 7 juvenile. The *Ci-Vasa*-positive cells (arrowheads) are entirely surrounded by somatic tissue. The number of *Ci-Vasa*-positive cells in the gonad increased. (Adapted from Shirae-Kurabayashi *et al.* (2006) *Development*, **133**, 2683–2693.) (n) Gonad formation in solitary ascidians. Vasa is segregated to the most posterior blastomeres during embryogenesis. The most posterior blastomeres become Vasa-expressing PGCs and form the gonadal rudiment together with somatic cells. It is unlikely that PGCs in the larval tail are the only germ cell progenitors. (Adapted from Kawamura *et al.* (2011) *Dev Dyn*, **240**, 299–308.)

As discussed in Chapter 5, B7.6 cells contain *postplasmic/PEM* mRNAs. With respect to the establishment of a germline, *postplasmic/PEM* mRNAs exhibit at least two different redistribution patterns during *C. intestinalis* tailbud embryo formation. B7.6 undergoes a supplementary asymmetric cell division during gastrulation to produce two distinct daughter cells, a smaller anterior cell, B8.11 (Fig. 14.1c, arrowhead) and a larger posterior cell, B8.12 (Fig. 14.1c, arrow). *Ci-pem* and *Ci-macho-1* mRNAs are segregated into B8.11 (Fig. 14.1f), and B8.11 derivates eventually associate with the juvenile gut wall, suggesting that B8.11 is not involved in germ cell formation. By contrast, *Ci-Vasa* (*Ci-VH*) mRNA and its protein are distributed to both B8.11 and B8.12 (Fig. 14.1b–e). Later, B8.12 divides and its progeny are incorporated into the juvenile gonad (Fig. 14.1g–m). Ci-Vasa protein is translated from its maternal mRNA source, and its expression is strongly upregulated in B8.12 descendants, leading to the formation of perinuclear Ci-Vasa granules, which are reminiscent of the electron dense "nuage" granules that characterize germ cells in many animal species.

The second redistribution pattern of *postplasmic/PEM* mRNAs includes those for *pem-3*, *ZF-1*, *GCNF*, and *POPK-1*. These factors are redistributed like *Ci-Vasa* mRNA into B7.6 and then B8.12. *pem-3* is the homolog of *mex-3*, which is associated with P granules in the *Caenorhabditis elegans* germline. Its maternal mRNA is distributed into B8.12 and could be involved in germ cell specification in a similar manner to *mex-3*. Therefore, *Ciona* serves as a comparatively straightforward system in which to investigate cellular and molecular mechanisms responsible for segregation of maternal factors, asymmetric division of germ- and somatic-line cells, and specification and subsequent differentiation of PGCs.

14.1.2   Induction of Primordial Germ Cells

Interestingly, cutting-off the posterior portion of the *Ciona* larval tail, which contains *Vasa*-positive cells, leads to the initial loss of PGCs in the developing juvenile gonad. However, these manipulated larvae eventually recover PGCs. Specifically, during later juvenile stages, germ cells again appear in the gonad. This rescue suggests that germ cells can originate from source(s) other than B7.6-derived PGCs. Although it is unclear whether this second source of germ cells also functions during normal development, at least under extreme circumstances, *Ciona* can derive germ cells by induction, as mammals do.

14.1.3   Repression of Somatic Gene Transcription in Germline Cells

In many animal embryos, germline progenitors actively repress somatic gene transcription, thereby ensuring separation of the germ and somatic lines. A common mechanism by which germline progenitors are protected from differentiation-inducing signals is a transient and global repression of ribonucleic acid polymerase II (RNAPII)-dependent transcription. This repression correlates with an absence of phosphorylation of Ser2 residues in the carboxy-terminal domain (CTD) of RNAPII, which is a critical modification for transcriptional elongation. In addition, Pgc in *Drosophila* and PIE-1 in *C. elegans* interact with positive transcription elongation factor b (P-TEFb), the CTD-Ser2 kinase complex, and they prevent P-TEFb recruitment to transcription sites.

In ascidian embryos, Pem protein is likely to act as Pgc and PIE-1. The Ci-Pem protein is localized to the germline blastomere nucleus and plays a pivotal role in repressing somatic gene transcription. An MO-mediated reduction in *Ci-pem* mRNA levels results in the ectopic expression of several somatic genes including those downstream of both maternal β-catenin and GATAa. Immunoprecipitation assays show that Ci-Pem protein interacts with two *Ciona* homologs of Groucho, a general corepressor of mRNA transcription. Ci-YB-1, one of the putative translational repressors associated with *post-plasmic/PEM* mRNAs, is also distributed to B8.11, and mediates repression of somatic gene transcription.

In *Halocynthia*, *pem* knockdown resulted in ectopic transcription and ectopic phosphorylation of CTD-Ser2 in the germline. Overexpression of *pem* abolished all transcription and led to the under phosphorylation of CTD-Ser2 in somatic cells. Pem protein was reiteratively detected in the nucleus of germline cells and coimmunoprecipitated with CDK9, a component of P-TEFb. The ascidian Pem therefore provides an example of evolutionary constraint on how a mechanism of germline silencing can evolve in diverse animals.

14.2   SPERMATOGENESIS

The ascidian testis is composed of male germ cells or sperm, testicular epithelium, and spermiduct (Fig. 14.1n). Development of the testis is first recognized by the formation of seminiferous tubes, and spermatogenesis proceeds in the tubule wall. Sperm of solitary ascidians are usually 50–60 μm long and consist of a short head (3–9 μm) and a long tail filamentous tail (48–53 μm), although they lack a middle piece (Fig. 14.2a). The tail contains a single axoneme exhibiting the usual 9 + 2 pattern of microtubules.

As discussed in Chapter 3, proteomics has recently been introduced into the *Ciona* system. For example, two-dimensional gel-electrophoresis analysis of the *Ciona* embryo detects ~600 main protein spots. There are few differences in the amount of major proteins among fertilized eggs, cleaved embryos, and tailbud embryos. However, some minor changes in the amount or isoelectric points of the proteins could be detected, suggesting that these proteins are likely to play roles in the cell cycle, cell differentiation, or morphogenesis.

On the other hand, proteomics provides a great amount of information about molecular components of ascidian sperm. For example, *Ciona* sperm were fractionated into heads and flagella, and then separated into Triton X-100-soluble and -insoluble fractions. Characterization of proteins from each fraction, as well as whole sperm by isoelectric focusing using two different pH ranges, followed by SDS-PAGE at two different polyacrylamide concentrations and mass spectrometry (MALDI-TOF), demonstrates ~300 nonredundant proteins specific to different sperm compartments. In addition, gene expression profiles in the testis or sperm have been determined. These studies reveal distinct localization of actin isoforms and novel Ca^{2+}-binding proteins in axonemes, the presence of G-protein-coupled signaling and ubiquitin pathways in sperm flagella, and localization of testis-specific serine/threonine kinase. These analyses also lead to the identification of important proteins with a minor presence in the cell, including the molecular motor units, the outer arm dynein, and the radial spoke (a substructure with a regulatory complex for dynein) from the axonemes of *Ciona* sperm flagella. In addition, recent studies reveal the functions of novel proteins, such as a dynein-related coiled-coil protein, a protein with a calmodulin-binding site and a ubiquitin-binding site (CMUB), and a novel MORN repeat protein, but also unexpected functions of well-known proteins, such as the heat shock protein HSP40 and nucleoside diphosphate kinase. Further studies using the *Ciona* sperm would facilitate exploration of the functions of uncharacterized sperm proteins and ultimately elucidate new molecular mechanisms in sperm physiology.

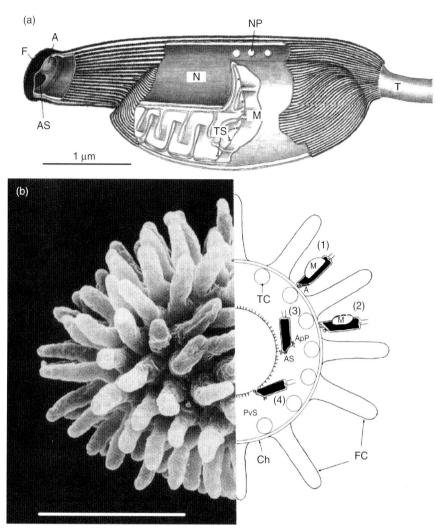

FIGURE 14.2   (a) Schematic illustration of *Ciona intestinalis* spermatozoon. An acrosome (A) is present at the anterior region of the head. AS, apical substance at the anterior tip of the head; F, fuzzy materials; M, mitochondrion; N, nucleus; NP, nuclear pore; T, tail; TS, tubular structure in the mitochondrion. (b) Representation of fertilization in *Ciona intestinalis*. The egg is surrounded by a chorion (Ch) or vitelline coat, to which spermatozoa approach through clefts between follicular cells (FC). (1) The spermatozoon binds to the surface of the chorion by the anterior tip of its head. (2) The acrosome reaction occurs on the surface of the chorion or after the penetration of the sperm apex through the chorion. The mitochondrion (M) is cast away during penetration of the chorion. (3) In the perivitelline space (PvS), apical processes (ApP) protrude from the apex of the sperm head. (4) Gamete fusion takes place between some of the apical processes and egg plasmalemma, resulting in the incorporation of the sperm from the anterior tip of its head. A, acrosome; AS, apical substance; TC, test cells. (Adapted from Satoh (1994) *Developmental Biology of Ascidians*, Cambridge University Press, New York.)

14.3 OOGENESIS

The aforementioned ascidian egg is enclosed in a tough, noncellular vitelline coat (VC), or chorion. On the outer surface of VC are attached many inner and outer follicular cells (FC). Within the perivitelline space between the egg and VC, there are many test cells. The existence of these test cells within the perivitelline space is a unique phenomenon in the animal kingdom. Oocyte growth of *C. intestinalis* is subdivided into four stages (Fig. 14.3). Stage I oocytes (50 μm in diameter) contain the smallest germinal vesicle (GV) and the smallest quantity of cytoplasm, and are surrounded by envelope organs consisting of undifferentiated primary FC (Fig. 14.3a). Stage II oocytes (50–70 μm in diameter) have prominently individualized cube-shaped follicle cells surrounding the oocytes (Fig. 14.3b). Stage III oocytes (100 μm in diameter) have more cytoplasm and more outstanding inner follicle structure (Fig. 14.3c), and automatically cause GVBD when exposed to seawater. Stage IV (after GVBD) oocytes, predominantly present in the oviduct, are capable of fertilization.

Tachykinins (TKs) are one of the largest families of neuropeptides found so far in vertebrates. In mammals, TKs and their receptors are expressed in the ovary, but the function of ovarian TKs remains to be elucidated. A recent study demonstrates that *C. intestinalis* contains prototypes of vertebrate TKs (Ci-TK-I) and their receptors (Ci-TK-R) and that TKs appear to play an essential role in oocyte growth (Fig. 14.3d). *In situ* hybridization and immunostaining demonstrate that *Ci-TK-I* is specifically expressed in the adult neural complex and Ci-TK-R in test cells residing in oocytes at the vitellogenic stage. DNA microarray and real-time PCR show that Ci-TK-I induces gene expression and subsequent enzymatic activities of several proteases, including cathepsin D, chymotrypsin, and carboxypeptidase B1, in the ovary. Treatment of *Ciona* oocytes with Ci-TK-I results in progression of their growth from the vitellogenic stage (stage II) to the post-vitellogenic stage (stage III). In addition, Ci-TK-I-induced oocyte growth is blocked by a TK antagonist or by protease inhibitors. These results indicate that Ci-TK-I enhances the growth of vitellogenic oocytes via upregulation of gene expression and protease activities, the first demonstration of a biological role, and the underlying molecular mechanism for a TK in the ovary.

In vertebrates, the hypothalamus produces gonadotropin-releasing hormone (GnRH), the anterior portion of the pituitary gland produces luteinizing hormone (LH) and follicle-stimulating hormone (FSH), and the gonads produce estrogen (ovary) and testosterone (testis). This hypothalamic–pituitary–gonadal axis (HPG axis) plays an essential role in reproduction including oocyte maturation. In addition, in oviparous vertebrates, the HPG axis is sometimes referred to as the hypothalamus–pituitary–gonadal–liver axis (HPGL-axis) since egg-yolk and chorionic proteins are synthesized heterologously in the liver, which are necessary for oocyte growth and development. Ascidians have no organs corresponding to the vertebrate pituitary gland, and GnRH is suggested to have nonreproductive function. *Ciona* has a prominent GnRH system spanning the entire length of the larval nervous system. The ascidian GnRH receptors are phylogenetically closest to vertebrate GnRH receptors, and functional analysis of the receptors shows that *Ciona* has evolved a unique GnRH system with multiple ligands and receptor heterodimerization enabling complex regulation. One of the GnRH genes is conspicuously expressed in larval motor ganglion and nerve cord, which are homologous structures to the hindbrain and spinal cord of vertebrates. Correspondingly, GnRH receptor genes are expressed in the tail muscle and notochord of embryos, both of which are phylotypic axial structures

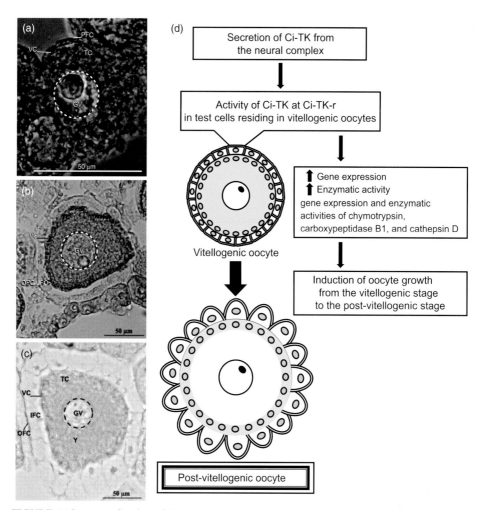

FIGURE 14.3   (a–c) Staging of *Ciona* oocytes during growth and maturation. Each stage is classified according to follicle structure and diameter. (a) Pre-vitellogenic (stage I) oocytes (50 μm in diameter). (b) Vitellogenic (stage II) oocyte (50–70 μm in diameter). (c) The post-vitellogenic (stage III) oocyte (100 μm m in diameter). GV, germinal vesicle; IFC, inner follicular cell; OFC, outer follicular cell; PFC, primary follicle cells; TC, test cell; VC, vitelline coat; Y, yolk. (d) Scheme of the essential molecular mechanism underlying the growth of vitellogenic oocytes triggered by Ci-TK-I. Ci-TK-I activates Ci-TK-R which is specifically expressed in test cells residing within vitellogenic oocytes, and which then upregulates gene expression and enzymatic activities of chymotrypsin, carboxypeptidase B1, and cathepsin D. Increased protease activities enhance the growth of oocytes from the vitellogenic stage (stage II) to the post-vitellogenic stage (stage III). (Adapted from Aoyama *et al.* (2008) *Endocrinology*, **149**, 4346–4356.)

along the nerve cord. There is evidence that GnRH-producing cells are present in the hindbrain and spinal cord of the medaka, *Oryzias latipes*, thereby suggesting the deep evolutionary origin of a nonreproductive GnRH system in chordates.

It is to ask whether the TK-dependent oocyte growth system is conserved in vertebrates. As the HPGL-axis is likely to be a vertebrate-specific innovation associated with later stages of oocyte growth, the TK-dependent oocyte growth system may represent a basic and/or more ancient system of oocyte growth, shared by chordates.

14.4 FERTILIZATION

The process of fertilization is schematically shown in Fig. 14.2b. Recent studies of the *Ciona* system have greatly advanced our understanding of molecular mechanisms of ascidian fertilization, especially in the field of sperm chemotaxis and self and nonself-recognition during fertilization.

14.4.1 The Sperm-Activating and -Attracting Factor

Chemotactic responses are important cell-to-cell communication systems. Chemotaxis of spermatozoa toward eggs during fertilization is observed in most animals and lower plants. In animals, sperm chemoattractants have been identified only in seven species, since remarkable species-specificity of ligand–receptor interaction makes it difficult to characterize these molecules. The chemoattractant of *C. intestinalis*, namely, the sperm-activating and -attracting factor (SAAF) has been revealed as a novel sulfated polyhydroxysterol, (25S)-3α,4β,7α,26-tetrahydroxy-5α-cholestane-3,26-disulfate. Optical isomerization does not affect SAAF activities. Hydrolysis at the sulfate group of SAAF decreases the sperm-activating and sperm-attracting activities, while hydrolysis on both sides results in loss of both activities. Biotinylated-SAAF loses its sperm-activating ability, but retains sperm-binding and chemotactic abilities. Thus, the sulfate groups of SAAF are responsible for these activities. The *C. intestinalis* SAAF acts as a SAAF for another species, *Ciona savignyi*.

Ca^{2+} is known to be important in sperm chemotaxis. A recent study showed that transient $[Ca^{2+}]_{(i)}$ increases in the flagellum (Ca^{2+} bursts) concomitantly with a change in the swimming direction in an SAAF gradient field. Ca^{2+} bursts induced by a local threshold SAAF concentration trigger a sequence of flagellar responses comprising quick turning followed by straight swimming to direct spermatozoa efficiently toward eggs.

14.4.2 Self- and Nonself-Recognition During Fertilization

14.4.2.1 *s-Themis and v-Themis, Proteins Involved in* Ciona *Sperm–Egg Self-Sterility*

A number of observations on sperm–egg self/nonself-discrimination have been made in *C. intestinalis*. The self/nonself-discrimination site on eggs resides in the VC, an acellular matrix surrounding the egg. The VC displays higher affinity to non-self than to self sperm, and a sperm's fertility is closely related to its ability to bind the VC. This suggests that the male self-incompatibility (SI) factor is expressed on the sperm surface, whereas the female SI factor is expressed on the VC. The self-sterility acquisition of VC takes place several hours after GVBD. Furthermore, the barrier against self-fertilization is abolished by treatment with mildly acidic seawater or proteases. Treating eggs with weak acid allows for artificial breeding of self-fertilized siblings, which produce a number of cross-sterile combinations, a rarely observed phenomenon in wild-type populations. Two types of cross-sterile combinations are observed: unidirectional and bidirectional. Unidirectional cross-sterility is unlikely to be due to a crossover between male and female SI genes, as such an event would disrupt the SI system after a few generations.

Based on these facts, Morgan (1944) introduced the "haploid sperm hypothesis" to explain SI (Fig. 14.4a). The theory proposes that SI specificity is determined by haploid gene expression in sperm, but by diploid gene expression in eggs. Specifically, a parent that is heterozygous at the SI locus (represented as A/a) will produce two populations of sperm (A-sperm and a-sperm), either of which can fertilize two types of homozygous eggs (A/A and a/a). By contrast, sperm (A-sperm and a-sperm) from homozygous individuals (A/A and a/a) would be unable to fertilize heterozygous eggs (A/a), since heterozygote eggs contain both types of female SI gene products at VC.

Nearly a century later, Harada *et al*. (2008) repeated Morgan's experiments and succeeded in isolating the candidate SI recognition genes (Fig. 14.4b). This achievement was realized by taking full advantage of the *C. intestinalis* draft genome sequences, transcriptomic and proteomic data, and a detailed physical map of the chromosomes. Two SI loci (designated as A and B) are involved in the self-sterility of *C. intestinalis* (Fig. 14.4b). Loci A and B are located on chromosomes 2q and 7q, respectively (Fig. 14.4b). Both loci contain a tightly linked pair of polycystin-1 receptor genes (*s-Themis-A* and *-B* and fibrinogen-like ligand genes (*v-Themis-A* and *-B*). The receptors (s-Themis-A and -B) and ligands (v-Themis-A and -B) are named after the Greek goddess of divine order, law, and custom, who prohibits incest. In both cases, the ligand gene is located in the first intron of its respective receptor gene and is encoded in the opposite direction. *s-/v-Themis-A* and *-B* are exclusive, extremely polymorphic genes, that are commonly encoded within the candidate regions of both loci. For example, only 29% of the amino acid sequence (42% of the nucleotide sequence) is identical between two alleles of *s-Themis-B*. In a model of the s-Themis-/v-Themis-mediated SI system that is controlled by two loci, each of the loci encodes both a sperm-side recognition molecule (s-Themis) and a VC-side recognition molecule (v-Themis). Both of the *Themis* loci have many alleles that comprise a single haplotype because of their extreme genetic proximity. s-Themis proteins may act as receptors that specifically interact with v-Themis proteins encoded in the same haplotype. When two s-Themis proteins on the sperm surface recognize their respective autologous v-Themis proteins as self on the VC, a sperm regards the egg as self, reducing its ability to bind to the VC.

s-Themis is expressed in the testis. Polycystin-1 (encoded by *PDK1*) is a calcium channel component protein that belongs to the transient receptor potential superfamily. Polycystin-1 participation in fertilization has been described in several other organisms, including sea urchins, in which it is proposed that PDK1 plays a role in the acrosomal reaction. The PKD1 signal is known to induce the elevated cytoplasmic $[Ca^{2+}]$ via the PKD2 channel. Sperm motility markedly decreases within 5 min after attachment to the VC of self eggs, but not after attachment to the VC of nonself eggs. Sperm detach from the self-VC or stop moving within 5 min after binding to the self-VC. Furthermore, sperm $[Ca^{2+}]_i$ rapidly and dramatically increases and is maintained at a high levels, in the head and flagellar regions when sperm interact with self-, but not nonself-VC. Therefore, it is highly likely that the sperm self-recognition signal triggers $[Ca^{2+}]_i$ increase and/or Ca^{2+} influx, which elicits a SI response to reject self-fertilization.

A more recent study identified an acid-extractable VC factor, named Ci-v-Themis-like, which has also the ability to distinguish self- from nonself-sperm. Ci-v-Themis-like is present in developing oocytes and on the VC in mature eggs. Yeast two-hybrid screenings revealed candidate proteins that interact with Ci-v-Themis-like, including sperm proteases and coiled-coil-domain-containing proteins. v-Themis-like and its binding partners are likely involved in sperm binding to the VC prior to the allorecognition process during

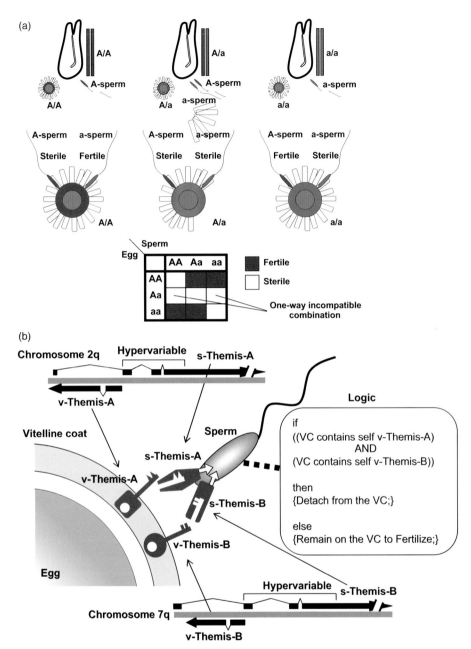

FIGURE 14.4 Self and nonself-recognition during *Ciona* fertilization. (a) The haploid sperm hypothesis proposed by Morgan to explain self-incompatibility (SI). (Upper panel) The double-line to the right of each animal represents the hypothetical diploid SI gene. Each of the homozygous parents (represented as A/A or a/a) produces haploid A-sperm or a-sperm, whereas the heterozygous parent produces two populations of sperm (A-sperm and a-sperm). Unidirectional cross-sterility is observed between the sperm of the homozygous parent and the egg of the heterozygous parent. (Lower panel) Owing to the presence of unidirectional cross-sterile combinations, the cross-sterility pattern forms a crank-like shape. (b) Schematic representation of the molecular mechanisms underlying the Themis-mediated SI system in *Ciona intestinalis*. (Adapted from Harada and Sawada (2008) *Int J Dev Biol*, **52**, 637–645.)

C. intestinalis fertilization. The molecular mechanisms involved in the ascidian self and nonself-recognition are likely to be similar to those that have been recently unveiled in plants. An interesting question is whether the *s-Themis/v-Themis* SI system functions in ascidians other than *Ciona*. Recent studies show that this system is involved in ascidian species such as *Halocynthia*, which release sperm and eggs simultaneously, but not in species such as *Styela*, which release sperm and eggs at different times such that sperm very rarely encounter self eggs. Accordingly, the genetic mechanisms involved in the evolutionary development of the *s-Themis/v-Themis* SI system may shed interesting light on topics in immunology.

14.4.2.2 Membrane Components Involved in Halocynthia Fertilization Self/Nonself-Recognition

In *H. roretzi*, other molecules are involved in sperm–egg binding. One is a 70 kDa VC protein, HrVC70, which consists of 12 epidermal growth factor (EGF)-like repeats. As nonself sperm rather than self sperm efficiently bind to HrVC70-agarose and since the attachment of this protein to VC during oocyte maturation is closely linked to self-sterility, HrVC70 is considered to be a candidate for the female SI factor. A binding partner of HrVC70, HrUrabin, is a 35 kDa glycosylphosphatidylinositol (GPI)-anchored glycoprotein in sperm lipid rafts. HrUrabin appears to play a key role in the binding of nonself sperm to HrVC70 during fertilization.

14.5   METAMORPHOSIS

Most metazoans have a biphasic life history. The fertilized egg develops first to a larva, which then metamorphoses to a juvenile and an adult. This indirect development has the advantage of providing metazoans two opportunities to invade new niches for feeding. Metamorphosis is therefore one of the most dynamic developmental processes, in which the larval form is partially destroyed to reconstruct the adult form. Ascidians provide excellent experimental systems to uncover cellular and molecular mechanisms involved in animal metamorphosis, since tadpole larvae, with a small number of constituent cells, exhibit metamorphosis synchronously within several days of its onset.

Metamorphosis of tadpole-like larvae to sessile juveniles and adults involves dynamic and complex changes in the shape and physiological function of constituent cells (Fig. 14.5a, b). During the swimming period, a larva obtains a capability termed competence to respond to a wide variety of external and endogenous signals for undergoing metamorphic changes. In *C. intestinalis*, competence requires ∼6 h. The preoral lobe, including the papilla, plays a role in metamorphosis initiation in response to the signals (Fig. 14.5a). Cloney (1982) pointed out the following 10 events as the basis of ascidian metamorphosis: (i) secretion of adhesives by the papillae or epidermis of the trunk, (ii) eversion and retraction of papillae, (iii) regression of the larval tail, (iv) loss of the outer cuticle layer of the larval tunic, (v) emigration of blood cells or pigmented cells, (vi) rotation of visceral organs through an arc of about 90°, expansion of the branchial basket, and elongation of the oozooids of juveniles, (vii) expansion, elongation, or reciprocation of ampullae, reorientation of test vesicles, and expansion of the tunic, (viii) retraction of the larval sensory vesicle, (ix) phagocytosis of the visceral ganglion, sensory organs, and cells of the axial complex, and (x) release of juvenile organ rudiments from an arrested state of development. These coordinated metamorphic events occur and proceed in the order enumerated above.

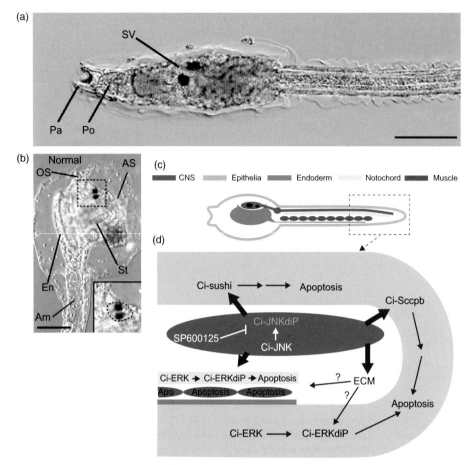

FIGURE 14.5 Metamorphosis of *Ciona intestinalis*. (a) A 3-dpf-old larva. Pa, papillae; Po, preoral lobe; SV, sensory vesicle. Bar, 100 μm. (b) A juvenile. Most adult organs are functional at this stage. Am, ampulla; AS, atrial siphon; En, endostyle; St, stigmata of pharyngeal gill; OS, oral siphon. Bar, 100 μm. Higher magnification of the sensory vesicle is shown in the inset. The outline of the sensory vesicle is visualized by a broken line. (c, d) A model of the role of the CNS in the regulation of apoptosis during metamorphosis. Ci-JNK activation in the CNS leads to *Ci-sushi* and *Ci-Sccpb* gene expression in epithelia. These genes are essential for initiating apoptosis at the onset of metamorphosis. Extracellular matrix (ECM) modification is also a result of Ci-JNK activation in the CNS, which could promote induction of apoptosis through Ci-ERK activation in adjacent tissues. (e) A possible classification of pathways of *Ciona* metamorphic events. Metamorphic events are classified based on the sensitivity to cellulose and *Ci-CesA*, and dependency on *trf*. The four groups of metamorphic events are shown by different colors. Cellulose-sensitive events are shown in orange, *trf*-dependent events in green, and cell division-required events in light blue. Retraction of papillae is categorized in both cellulose-sensitive events and *trf*-dependent events. *Ci-CesA* and/or cellulose are likely to be involved in the suppression of the cellulose-sensitive events. The causal gene of the *trf* mutant is involved in a metamorphic pathway, tentatively called the "*trf*-pathway". A metamorphic pathway activates cell division in adult organ primordia, and initiates cell division-required events. (a, b, e adapted from Nakayama *et al.* (2009) *Dev Biol*, **326**, 357–367; c, d from Chambon *et al.* (2007) *Development*, **134**, 1203–1219.)

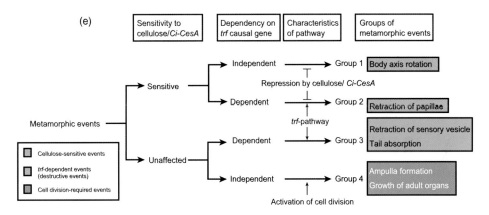

FIGURE 14.5   (*Continued*)

Recent studies demonstrate that the MAPK, c-JUN N-terminal kinase (JNK) and ERK, are both required for *Ciona* metamorphosis (Fig. 14.5c, d). JNK cascade is activated at the time of competence in the tadpole CNS and the pharyngeal rudiment. Inhibition of the JNK pathway completely blocks metamorphosis. Similarly, ERK activation correlates with the time of metamorphic competence. ERK is activated in the tail muscle, the notochord, the cytoplasm of proximal palp cells, the anterior and posterior sensory vesicle, the epidermis overlying the sensory vesicle, the neck region, atrial primordia, and in the epidermis of the tail. As all these larval organs and tissues degenerate during metamorphosis, this suggests that the phosphorylated form of ERK transduces the death-activating signal in tail tissues during metamorphosis. Blocking of ERK activation inhibits metamorphosis in a significant percentage of tadpole larvae. Microarray analysis reveals that either ERK or JNK activity controls a network of genes, which is involved in hormonal signaling, in innate immunity, in cell–cell communication and in the extracellular matrix. Of these, Ci-sushi, a cell–cell communication protein controlled by JNK activity, is required for the wave of apoptosis that precedes tail regression. A model of metamorphosis is proposed whereby JNK activity in the CNS induces apoptosis in several adjacent tissues that compose the tail, by inducing expression of genes such as *Ci-sushi* (Fig. 14.5d).

Another question is how various elementary events of metamorphosis are coordinated. This question is examined using larvae from which the papillae and preoral lobes are cut off, those from which the tail is cut off, and those of two mutants, *swimming juveniles* (*sj*) (see Section 8.3) and *tail-regression fail* (*trf*) of *C. intestinalis* (Fig. 14.5e). Papillae-cut larvae initiate certain trunk metamorphic events such as the formation of an ampulla, body axis rotation, and adult organ growth without other metamorphic events, suggesting that metamorphic events can be divided into at least two groups, events initiated, and not initiated in papillae-cut larvae.

Although the causal gene of *trf* remains to be characterized, *trf* larvae show phenotypes similar to those of papillae-cut larvae. By comparing the phenotypes of *sj*, *trf*, and papillae-cut larvae, metamorphic events of *C. intestinalis* can be divided into the following four groups (Fig. 14.5e). Group 1 comprises a cellulose-sensitive and *trf*-independent event, body axis rotation. Group 2 is a cellulose-sensitive and *trf*-dependent event, papillae retraction. Group 3 comprises cellulose-independent and *trf*-dependent events, including sensory vesicle retraction and tail absorption. Group 4 comprises cellulose-independent and *trf*-independent events, including ampulla formation and adult organ growth. Interestingly, events belonging to the same group have common characteristics. As mentioned above, *trf*-dependent events, which include group 2 and 3 events, are possibly dependent on the nervous system. *trf*-dependent events have another common characteristic; they are all events that destroy the larval body. Group 4 events require cell division and these events are directly involved in formation of adult bodies, called "cell division-required events." It is likely that events within a given group are regulated by the same pathway.

SELECTED REFERENCES

AOYAMA, M., KAWADA, T., FUJIE, M., HOTTA, K. *et al.* (2008) A novel biological role of tachykinins as an up-regulator of oocyte growth: identification of an evolutionary origin of tachykiniergic functions in the ovary of the ascidian *Ciona intestinalis*. *Endocrinology*, **149**, 4346–4356.

CHAMBON, J.P., NAKAYAMA, A., TAKAMURA, K. McDOUGALL, A. *et al.* (2007) ERK- and JNK-signalling regulate gene networks that stimulate metamorphosis and apoptosis in tail tissues of ascidian tadpoles. *Development*, **134**, 1203–1219.

CLONEY, R.A. (1982) Ascidian larvae and the events of metamorphosis. *Am Zool*, **22**, 817–826.

HARADA, Y., SAWADA, H. (2008) Allorecognition mechanisms during ascidian fertilization. *Int J Dev Biol*, **52**, 637–645.

HARADA, Y., TAKAGAKI, Y., SUNAGAWA, Y., SAITO, T. *et al.* (2008) Mechanism of self-sterility in a hermaphroditic chordate. *Science*, **320**, 548–550.

INABA, K., NOMURA, M., NAKAJIMA, A., HOZUMI, A. (2007) Functional proteomics in *Ciona intestinalis*: a breakthrough in the exploration of the molecular and cellular mechanism of ascidian development. *Dev Dyn*, **236**, 1782–1789.

KAWAMURA, K., TIOZZO, S., MANNI, L., SUNANAGA, T. *et al.* (2011) Germline cell formation and gonad regeneration in solitary and colonial ascidians. *Dev Dyn*, **240**, 299–308.

KUMANO, K., TAKATORI, N., NEGISHI, T., TAKADA, T. *et al.* (2011) A maternal factor unique to ascidians silences the germline via binding to P-TEFb and RNAP II regulation. *Curr Biol*, **21**, 1308–1313.

KUSAKABE, T.G., SAKAI, T., AOYAMA, M., KITAJIMA, Y. *et al.* (2012) A conserved non-reproductive GnRH system in chordates. *PLoS One*, **7**, e41955.

MORGAN, T.H. (1944) The genetic and phylogenetical problems of self-sterility in *Ciona*. VI. Theoretical discussion of genetic data. *J Exp Zool*, **95**, 37–59.

NAKAYAMA, A., CHAMBON, J.-P., HORIE, T., SATOH, N. *et al.* (2009) Delineating metamorphic pathways in the ascidian *Ciona intestinalis*. *Dev Biol*, **326**, 357–367.

OTSUKA, K., YAMADA, L., SAWADA, H. (2013) cDNA Cloning, localization and candidate binding partners of acidextractable vitelline-coat protein Ci-v-Themis-like in the ascidian *Ciona intestinalis*. *Mol Reprod Dev*, doi: 10.1002/mrd.22213.

SAITO, T., SHIBA, K., INABA, K., YAMADA, L. *et al.* (2012) Self-incompatibility response induced by calcium increase in sperm of the ascidian *Ciona intestinalis*. *Proc Natl Acad Sci USA*, **109**, 4158–4162.

SATOH, N. (1994) *Developmental Biology of Ascidians*, Cambridge University Press, New York.

SAWADA, H., TANAKA, E., BAN, S., YAMASAKI, C. *et al.* (2004) Self/nonself recognition in ascidian fertilization: vitelline coat protein HrVC70 is a candidate allorecognition molecule. *Proc Natl Acad Sci USA*, **101**, 15615–15620.

SHIBA, K., BABA, S.A., INOUE, T., YOSHIDA, M. *et al.* (2008) Ca^{2+} bursts occur around a local minimal concentration of attractant and trigger sperm chemotactic response. *Proc Natl Acad Sci USA*, **101**, 15615–15620.

SHIRAE-KURABAYASHI, M., NISHIKATA, T., TAKAMURA, K., TANAKA, K.J. *et al.* (2006) Dynamic redistribution of vasa homolog and exclusion of somatic cell determinants during germ cell specification in *Ciona intestinalis*. *Development*, **133**, 2683–2693.

SHIRAE-KURABAYASHI, M., MATSUDA, K., NAKAMURA, A. (2011) Ci-Pem-1 localizes to the nucleus and represses somatic gene transcription in the germline of *Ciona intestinalis* embryos. *Development*, **138**, 2871–2881.

TAKAMURA, K., FUJIMURA, M., YAMAGUCHI, Y. (2002) Primordial germ cells originate from the endodermal strand cells in the ascidian *Ciona intestinalis*. *Dev Genes Evol*, **212**, 11–18.

YOSHIDA, M., MURATA, M., INABA, K., MORISAWA, M. (2002) A chemoattractant for ascidian spermatozoa is a sulfated steroid. *Proc Natl Acad Sci USA*, **99**, 14831–14836.

CHAPTER 15

INNATE IMMUNE SYSTEM AND BLOOD CELLS

The ability to discriminate self from nonself, also called allorecognition, is essential for individuals to survive under various conditions, and therefore self-/nonself-recognition systems are present in a wide variety of metazoans including sponges, cnidarians, echinoderms, ascidians, and vertebrates. In vertebrates, the adaptive immune system is responsible for allorecognition, in which the highly polymorphic major histocompatibility complex (MHC) proteins are coupled to immune receptor recognition. However, no evidence for the presence of adaptive immunity has been reported from invertebrates thus far. On the other hand, all metazoans have various types of innate immunity, some common with vertebrates and others specific to certain groups. To understand the origin and evolution of a vertebrate immune system, it is essential to understand the immune system of the closer relatives of vertebrates, namely tunicates. This chapter discusses innate immunity in ascidians, and blood cells and their special functions as well.

15.1 INNATE IMMUNITY IN *Ciona*

With the sequencing of the *Ciona intestinalis* genome, a comprehensive picture of its immune system-related genes has been obtained with comparison of vertebrate genomes. The pivotal genes for adaptive immunity, for example, MHC class I and II genes, T-cell receptors, and dimeric immunoglobulin molecules, have not been identified in the *Ciona* genome. By contrast, many genes involved in innate immunity have been identified, including Toll-like receptors (TLRs), complement components, and genes involved in intracellular signal transduction of immune responses. Collectively, these genes appear to be expanded and unexpectedly diverse in comparison with their vertebrate homologs.

15.1.1 Toll-Like Receptors

TLRs play pivotal roles in host defenses via the innate immune system. All TLRs are type I transmembrane proteins that harbor an intracellular Toll/interleukin-1 receptor (TIR) domain and extracellular leucine-rich repeat (LRR) motifs. LRRs exhibit specific pathogenic ligand recognition, and TIR participates in the activation of downstream signaling pathways. Nine functional TLRs (hTLRs) have been identified in humans, and each hTLR directly recognizes specific ligands (or pathogen-associated molecular

Developmental Genomics of Ascidians, First Edition. Noriyuki Satoh.
© 2014 John Wiley & Sons, Inc. Published 2014 by John Wiley & Sons, Inc.

patterns; PAMPs). TLRs or their related genes have been found in various metazoans. Amphioxus and sea urchin possess a great number of TLRs or TLR-related genes: 72 genes in amphioxus and 222 in sea urchin. Molecular phylogeny demonstrates that most TLR genes have been generated via species-specific gene duplication. In contrast, a *Ciona* genome survey and molecular cloning demonstrate the presence of only two TLRs in *C. intestinalis*, namely, Ci-TLR1 and Ci-TLR2. Both are typical TLRs composed of TIR, transmembrane, and LRR domains, although 7 and 13 LRRs are found in Ci-TLR1 and Ci-TLR2, respectively.

Ci-tlr1 and *Ci-tlr2* are abundantly expressed in the stomach, anterior and middle intestine (but not in posterior intestine), numerous hemocytes and, to a lesser degree, in the central nervous system, suggesting that Ci-TLRs function mainly in the alimentary tracts and hemocytes. Ci-TLR1 and Ci-TLR2 activate NF-κB in response to multiple TLR ligands, which are recognized by different mammalian TLRs. Reflecting the intestinal expression profile, the upregulation of NF-κB is found only in the stomach and middle intestine, but not in the posterior intestine. Interestingly, unlike any other vertebrate TLRs, both Ci-TLRs are present on both the plasma membrane and a number of late endosomes. These results lead to two important suggestions. Firstly, Ci-TLRs, like vertebrate TLRs, directly recognize their PAMPs and trigger the transactivation of NF-κB. Secondly, Ci-TLRs are "functionally hybrid TLRs" of vertebrate cell-surface TLRs and endosomal TLRs. Elucidation of PAMPs and intracellular localization of sea urchin, amphioxus, and cyclostome TLRs contributes not only to understanding of their biological roles but also molecular and functional divergence of the invertebrate TLR family.

15.1.2 Complement System

The mammalian complement system consists of more than 30 plasma and cell-surface proteins interacting in the recognition and elimination of pathogens. Three major physiological functions of the system are opsonization of the foreign particles, induction of inflammatory reactions, and cytolysis. The origin of the multicomponent complement system consisting of C3, Bf (factor B), and MASP (mannan-binding lectin associated serine protease) is traced back to the common ancestor of Eumetazoa. Both *Ciona* and *Halocynthia* have genes for C3, Bf, MASPs, mannan-binding lectin (MBL), ficolin, and CR3α and CR3β (Fig. 15.1a). *Halocynthia* C3, ficolin, and GBL proteins isolated from the body fluid are shown to act as a component of the opsonic complement system. In addition, the C3α fragment of *Ciona* C3 has chemotactic activity, indicating a possible role of the complement system in inflammation. These results show that, among the three major physiological functions of the system, the former two are conserved between mammals and urochordates.

In contrast, the third activity of the mammalian complement system, cytolytic activity, has not been recognized in the urochordate complement system. Although there are several C6-like genes with the membrane attack complex/perforin (MACP) domain in the *C. intestinalis* genome, all of them lack the C-terminal short consensus repeat (SCR) and factor I/membrane attack complex (FIM) domains, reported to be essential for interaction with other complement components (Fig. 15.1a). Thus, it is unlikely that these C6-like molecules are integrated in the urochordate complement system. All these results indicate that the urochordate complement system represents the primitive evolutionary stage just before the development occurred in the common ancestor of vertebrates.

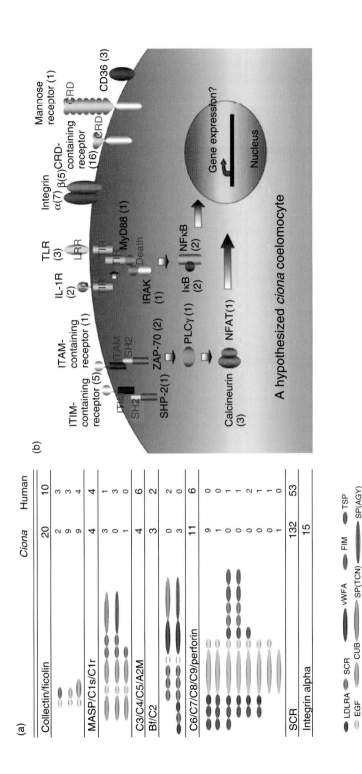

FIGURE 15.1 Innate immunity-related genes in the *Ciona intestinalis* genome. (a) Possible complement genes. Only genes with the same or nearly the same domain structures as those found in their vertebrate counterparts are counted. The left half of the figure shows the schematic domain structures of complement genes, and the right half indicates the copy numbers of those genes observed in *Ciona* and *Homo*. Domain identities are explained at the bottom. (b) Innate immunity-related transmembrane proteins and signaling molecules. Genes are placed in the figure according to the presumed cellular localization of their protein products. Arrows connecting individual components are placed according to the assumption that *Ciona* and mammals share basic signal transduction pathways. Identities of effector genes involved in cell proliferation and differentiation, inflammation, and antimicrobial responses are not known. Names of each gene and domain are shown in black and blue, respectively. The number of gene models identified is indicated in parentheses for each gene. (Adapted from Azumi et al. (2003) *Immunogenomics*, **55**, 570–581.)

In addition, a number of *Ciona* genes are predicted to encode integral membrane proteins with extracellular C-type lectin or immunoglobulin domains and intracellular immunoreceptor tyrosine-based inhibitory motifs (ITIMs) or immunoreceptor tyrosine-based activation motifs (ITAMs) along with their associated signal transduction molecules (Fig. 15.1b). A crucial component of vertebrate adaptive immunity is somatic diversification; however, recombination activating genes (RAGs) and activation-induced cytidine deaminase (AID) genes responsible for the generation of diversity are not present in *Ciona*. However, there are key V regions, the essential feature of an immunoglobulin superfamily VC1-like core, and possible proto-MHC regions scatter throughout the *Ciona* genome.

15.2 BLOOD CELLS

As ascidians are the invertebrate chordates closest to vertebrates, their blood cells have attracted researchers in the field of innate immune systems. Recent electron and light microscopic studies, as well as gene expression profiles, demonstrate at least eight types of hemocytes, which are distinguishable as two main types (Fig. 15.2). The first group includes agranular hemocytes such as hemoblasts, circulating lymphocyte-like cells (LLC), and hyaline amebocytes. The second group includes granular hemocytes such as granulocytes with small granules, granulocytes with large granules, unilocular refractile granulocytes (URG), and morula cells. Compartment cells and signet ring cells could be intermediate or terminal states, presumably involved in releasing inflammatory factors or tunic matrix components (Fig. 15.2).

15.3 VANADIUM CONCENTRATION

In 1911, the German physiologist, Martin Henze discovered high levels of vanadium accumulation in ascidian blood cells. This finding not only triggered vanadium research but also aroused great interest in the question of how such extraordinarily high levels of vanadium could be accumulated and what the role of vanadium in ascidians could possibly be. Recent approaches from inorganic, catalytic, and applied chemistry, as well as physiological, molecular, and pharmaceutical biology, have greatly advanced our understanding of this unique biological phenomenon.

Several ascidians such as *Ascidia gemmata* accumulate vanadium at concentration as high as 350 mM, corresponding to 10^7 times the level in seawater (35 nM), in vanadocytes or signet ring cells (Fig. 15.2 and Fig. 15.3a). This is thought to be the highest rate of accumulation of a metal in any living organism. In seawater, vanadium is dissolved in the +5 oxidation state (HVO_4^{2-} or $H_2VO_4^-$; V^V) (Fig. 15.3a). During the accumulation process, V^V is reduced to the +3 oxidation state (V^{3+}; V^{III}) via the +4 oxidation state (VO^{2+}; V^{IV}) (Fig. 15.3a). NADPH is a strong candidate to participate in reduction of V^V to V^{IV} with the assistance of several chelating substances. Enzymes involved in the pentose phosphate pathway are expressed exclusively in the cytoplasm of vanadocytes. In addition, Michibata's group has successfully isolated vanadium-binding proteins named Vanabins, Vanabin1−4, and VanabinP, from vanadocytes as well as blood plasma (Fig. 15.3b). Vanabins are capable of binding as many as circa 10−20 V^{IV} ions per molecule of protein and are thought to play an important role in V^V-reduction, V^{IV}-transport, and V^{IV}-storage processes in vanadocytes (Fig. 15.3a).

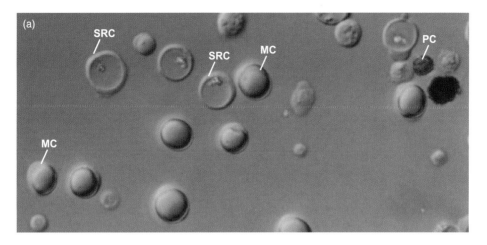

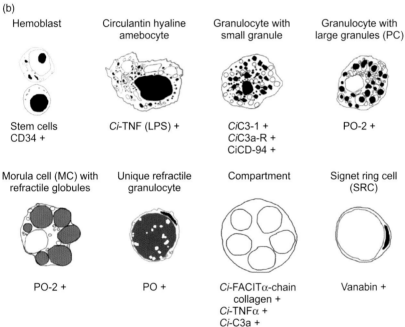

FIGURE 15.2 (a) Photograph of blood cells of *Ciona intestinalis*. MC, morula cells; PC, pigment cells; SRC, signet ring cells. (By courtesy of Ueki *et al*. (2003).) (b) Representative features, activities, and innate immunity gene expression of *C. intestinalis* inflammatory hemocytes. Hemocyte drawings are based on light and electron microscographs. (Modified from Arizza and Parrinello (2009) *Invertebr Survival J*, **6**, S58–S66.)

The content of the vanadocyte vacuole is maintained at an extremely low pH ($\sim$1.9), and the concentration of vanadium is correlated with proton concentration. A vacuolar-type H^+-ATPase (V-ATPase) is localized on the vacuolar membrane of vanadocytes and V-ATPase maintains low pH in vanadocyte vacuoles. It is hypothesized that a proton gradient generated by V-ATPase provides the energy to transport vanadium across the vacuolar membrane.

The Nramp (natural resistance-associated macrophage protein)/DCT1 family is known to transport a broad range of divalent cations (Fe^{2+}, Cu^{2+}, Zn^{2+}, Mn^{2+}, Cd^{2+}, Co^{2+}, Ni^{2+}, and Pb^{2+}) across the membrane using a proton gradient as the motive force. A cDNA encoding a member of the Nramp/DCT1 family was cloned from vanadocytes of *Ascidia sydneiensis samea*. The gene product, *As*Nramp, is localized on the vacuolar membrane and can operate as an antiporter of V^{IV} and H^+.

Further interdisciplinary studies might facilitate our understanding of the high levels of vanadium, identification of vanadium-accumulating blood cells, energetics of vanadium accumulation, the sulfate transport system, the redox mechanism of vanadium, and possible physiological roles of vanadium in ascidians.

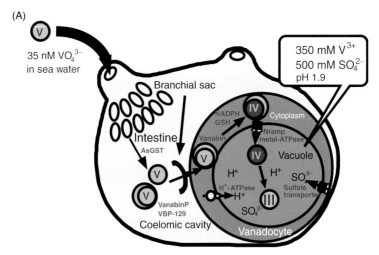

FIGURE 15.3   Vanadium accumulation in ascidian blood cells. (A) Schematic representation of molecular mechanisms of vanadium accumulation and reduction. The concentration of vanadium in the +5 oxidation state is only 35 nM in seawater. On the other hand, the highest concentration of vanadium in ascidian blood cells is 350 mM, and the concentration of sulfate is 500 mM. The vacuole interior is maintained at an extremely low pH (1.9) by H^+-ATPases. In this environment, almost all of the vanadium accumulated is reduced to V^{III} via V^{IV}. The first step in vanadium uptake may occur at a branchial sac or intestine, where glutathione-S-transferase functions as a major vanadium carrier protein. Then, vanadium-binding proteins, Vanabins, carry vanadium in the blood plasma and cytoplasm of vanadocytes or signet ring cells. The pentose phosphate pathway in the blood cell cytoplasm produces NADPH, which reduces V^V to V^{IV}. A metal-ATPase that might be involved in vanadium transport is present in the vacuolar membrane. (B) Structure of Vanabin2. (a) Amino acid sequence of Vanabin2. Disulfide bond pairings are indicated at the top of the sequence. Secondary structure elements of Vanabin2 are indicated at the bottom of the sequence and are colored correspondingly in all panels. (b) The final 10 structures superposed over the backbone heavy atoms of residues 18–70. Side chains of half-cysteine residues are shown as yellow lines. (c) Ribbon representation of a single structure in the same orientation as in panel b. (Adapted from Michibata and Ueki (2011), in *Vanadium: Biochemical and Molecular Biological Approaches* (ed. H. Michibata), Springer, UK.)

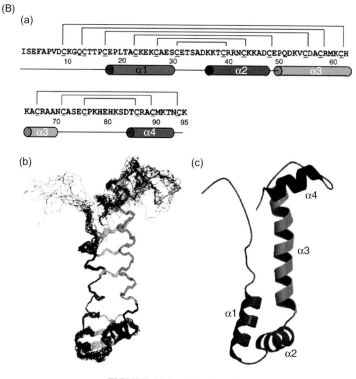

FIGURE 15.3   (*Continued*)

SELECTED REFERENCES

ARIZZA, V., PARRINELLO, D. (2009) Inflammatory hemocytes in *Ciona intestinalis* innate immune response. *Invertebr Survival J*, **6**, S58–S66.

AZUMI, K., DE SANTIS, R., DE TOMASO, A., RIGOUTSOS, I. *et al.* (2003) Genomic analysis of immunity in a Urochordate and the emergence of the vertebrate immune system: "waiting for Godot". *Immunogenomics*, **55**, 570–581.

MICHIBATA, H., UEKI, T. (2011) High levels of vanadium in ascidians, in *Vanadium: Biochemical and Molecular Biological Approaches* (ed. H. Michibata), Springer, UK.

NONAKA, M. (2001) Evolution of the complement system. *Curr Opin Immunol*, **13**, 69–73.

SATAKE, H., SEKIGUCHI, T. (2012) Toll-like receptors of deuterostome invertebrates. *Front Immunol*, **3**, 34.

UEKI, T., ADACHI, T., KAWANO, S., AOSHIMA, M. *et al.* (2003) Vanadium-binding proteins (vanabins) from a vanadium-rich ascidian *Ascidia sydneiensis samea. Biochim Biophys Acta*, **1626**, 43–50.

CHAPTER 16

COLONIAL ASCIDIANS: ASEXUAL REPRODUCTION AND COLONY SPECIFICITY

While many ascidians live as individuals (solitary or simple ascidians), others form colonies (colonial or compound ascidians). The colonial life style evolved several times independently in various ascidian orders (Fig. 1.2). Colonial ascidians propagate via asexual reproduction and regeneration, where stem-like (totipotent, undifferentiated) cells play pivotal roles in transmitting genetic information to the next generation. This chapter discusses developmental biology of colonial ascidians.

16.1 ASEXUAL REPRODUCTION OF COLONIAL ASCIDIANS AND STEM CELLS IN THE BUDDING

16.1.1 Asexual Reproduction

Colonial ascidians such as *Botryllus*, *Botrylloides*, and *Polyandrocarpa* reproduce both sexually and asexually, by budding and by strobilation (Fig. 16.1A). Extensive diversity exists with respect to the modes of bud formation, which is generally classified into two functional types—propagative and survival. Propagative budding accounts for colony growth, while survival budding is passive and serves primarily to insure the animal's survival during adverse environmental conditions. However, regardless of the mode of budding, the oozooid (zooid developed from an egg) usually reproduces only asexually, and does not attain sexual maturity. On the other hand, the blastozooid (zooid developed from a bud) reproduces both sexually and asexually. In other words, without at least one phase of asexual development, colonial ascidians would not propagate sexually, suggesting that asexual reproduction is an adaptation for propagation in colonial ascidians. Therefore, disclosure of gene regulatory networks that are responsible for the state transition from oozooids to blastozooids is an interesting subject for future research.

16.1.2 Bud Formation

Modes of bud formation can be categorized based on cell types from which major organs of the bud are derived (Fig. 16.1). That is, although a bud originates from a part of the parent body, not every cell type constituting the parent body is necessarily

Developmental Genomics of Ascidians, First Edition. Noriyuki Satoh.
© 2014 John Wiley & Sons, Inc. Published 2014 by John Wiley & Sons, Inc.

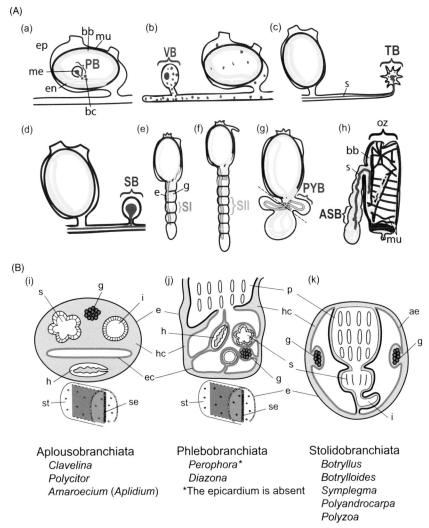

Aplousobranchiata
Clavelina
Polycitor
Amaroecium (Aplidium)

Phlebobranchiata
*Perophora**
Diazona
*The epicardium is absent

Stolidobranchiata
Botryllus
Botrylloides
Symplegma
Polyandrocarpa
Polyzoa

FIGURE 16.1   (A) Budding types of colonial ascidians, classified according to the position of the bud and the origin of germ layers (blue, ectoderm; red, mesoderm; orange, mesendodermal tissues; yellow, endoderm; and gold, epicardium). (a) Peribranchial budding (PB). (b) Vascular budding (VB). (c) Terminal budding (TB). The septum (s), an epithelial sheet within the lumen of the extra-zooidal vasculature, is shown. (d) Stolonial budding (SB). (e) Type I strobilation of the abdomen (SI). (f) Type II strobilation of abdomen and postabdomen (SII). (g) Pyloric budding (PYB). A dashed line separates the two precursor zooids that result from pyloric budding. (h) Aggregate stolonial budding (ASB); the stolon-like structure is shown. Bb, branchial basket; bc, blood cell; e, epicardium; en, endostyle; ep, epidermis; g, gut; me, mesendoderm; mu, muscle; oz, oozoid; s, septum. (Adapted from Brown and Swalla (2012) *Dev Biol*, **369**, 151–162.) (B) Schematic illustration of a zooidal organization with reference to multipotent cells (red) in colonial tunicates. (i) Transverse section of the abdomen in Aplousobranchiata. All species possess multipotent epicardial cells. In *Clavelina*, septum cells are also multipotent. (j) Longitudinal section of a zooid in Phlebobranchiata. The epicardium is present in all species, except *Perophora*. In *Perophora*, the septum is multipotent. (k) Frontal section of a zooid in Stolidobranchiata. The peribranchial (atrial) epithelium is multipotent. ae, atrial epithelium; e, epidermis; ec, epicardium; g, gonad; h, heart; hc hemocoel; i, intestine; p, pharynx; s, stomach; se, septum; st, stolon. (Adapted from Kawamura *et al.* (2008) *Dev Growth Diff*, **50**, 1–11.)

incorporated into a bud. Therefore, unincorporated cell types are supplied by multipotent cells that already exist in the bud. Such multipotent cells are those of the epicardial epithelium (Fig. 16.1i, j), septal mesenchyme (Fig. 16.1i, j), hemoblasts (Fig. 16.1j, k), and peribranchial epithelium (Fig. 16.1k). Usually, a bud contains at least one of these multipotent cells. In epicardial budding, the epicardial epithelium plays a central role. This type of budding includes strobilation of the abdomen (*Clavelina* and others) (Fig. 16.1e), the postabdomen (*Ritterella* and others), both the abdomen and postabdomen (*Polyclinum* and others) (Fig. 16.1f), and bud formation from the epicardial epithelium of the parent and epidermis (*Distaplia* and others) (Fig. 16.1a). In stolonial budding (Fig. 16.1d), the septal mesenchyme and hemoblasts are the major contributors to bud formation, as in *Clavelina*, *Perophora*, and others. In peribranchial budding, buds are formed via protrusion of the parent zooid body wall (Fig. 16.1a). This mode of budding is found in *Botryllus*, *Botrylloides*, and others.

All three germ layers are involved in bud formation. The ectoblast (ectoderm) of a new zooid is derived exclusively from the epidermis of the mother zooid, and the mesoblast (mesoderm) of the mother zooid is usually inherited by a new zooid. By contrast, the origin of the endoblast (endoderm) varies among colonial ascidians (Fig. 16.1B). In the family Clavelininae, for example, the endoblast originates from mesodermal mesenchyme. In the Polyclinidae, the endoblast is derived from endodermal epicardial epithelium, whereas in the Botryllidae the endoblast usually originates from ectodermal peribranchial epithelium. In addition, in some *Botryllus* and *Botrylloides* species, the endoblast of the vascular bud is derived from hemoblasts of mesodermal origin. Similarly, the origins of certain organs also vary among ascidians. For example, the neural gland is formed from septa-1 mesenchyme of the stolon in the Clavelininae, from endodermal epicardial epithelium in the Polyclinidae, and from ectodermal peribranchial epithelium in the Botryllidae. In other words, asexual reproduction involves a variety of cell and tissue sources. As mentioned above, in *Polyandrocarpa misakiensis*, the peribranchial epithelium shows developmental multipotency. The transdifferentiation of the multipotent epithelium requires at least one cycle of cell division and is likely to be triggered by endogenous RA. RA acts on mesenchymal cells, which then secrete proteases that serve as an actual transdifferentiation factor of peribranchial epithelium.

Hemoblasts or coelomic cells of colonial ascidians have the potential to give rise to several different types of cells/tissues. In *P. misakiensis*, hemoblasts or lymphocytes play a significant role in the development of body wall smooth muscle cells and in rudimentary gut formation. Prior to gut formation, many hemoblasts aggregate at the site where the rudiment develops. In addition, hemoblasts give rise to germ cells in colonial ascidians.

16.2   GERM CELL FORMATION IN COLONIAL ASCIDIANS

Morphology of the ascidian gonad is similar between solitary and colonial species. The testis consists of a variable number of testicular follicles, and the ovary consists of ovarian tubes that are thickened forming the germinal epithelium with stem cells as female germ cells (Fig. 16.2B). Peculiar accessory cells that are of germline origin, accompany the oocytes.

In colonial ascidians, such as *Botryllus primigenus* and *Botrylloides violaceus*, *Vasa* is expressed strongly by germline cells in the gonad. In *Ciona*, *Vasa*-expressing germline cells in the gonad can be traced back to B7.6, the most posterior blastomere in the

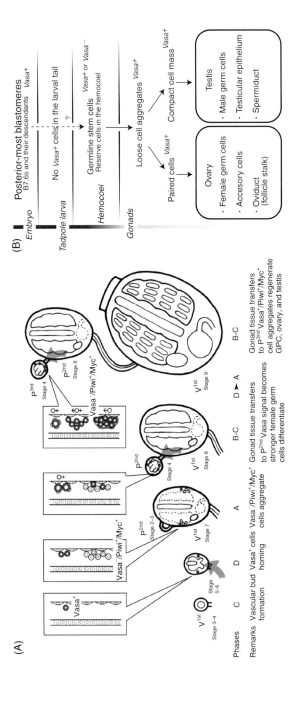

FIGURE 16.2 (A) A schematic representation of gonad formation during asexual reproduction of the colonial ascidian, *Botryllus primigenus*. In colonial phase C, vascular buds (V[1st]) become double-walled vesicles (stages 3–4). In phase D, a few Vasa[+] cells (red circles) extend from the test vessel. In phase A, Piwi[+]/Myc[+] cells aggregate just behind the zooid bud primordium (V[1st], stage 7). They are Vasa[−] or Vasa[+/−] (colorless and pinkish circles). In phase B, Vasa[+] cells increase in number and enter the newly formed palleal bud (P[2nd]). They begin to differentiate into female germ cells. The cell aggregate in p[2nd] has stronger Vasa signals. In phases B–C of the next blastogenic cycle, the third generation bud (P[3rd]) contains a set of Vasa[+]/Piwi[+]/Myc[+] gonadal components, including germline precursor cells, primordial testis, and primordial ovary (oogonia and oocytes). (Adapted from Kawamura and Sunanaga (2011) *Mech Dev*, **128**, 457–470.) (B) Gonad formation in colonial ascidians. Vasa is segregated to the posterior-most blastomeres during embryogenesis. However, *Vasa*-expressing cells disappear during larval formation. Germline competent cells reappear in the hemocoel of adult colonies. The latter germline cells are likely to be responsible for the flexibility of gonad formation. (Adapted from Kawamura *et al.* (2011) *Dev Dyn*, **240**, 299–308.)

112-cell embryo (Fig. 14.1). However, *Vasa*-positive cells disappear after the tailbud embryo stage of colonial ascidians, obscuring the connection of embryonic germline cells to PGCs in the adult gonad (Fig. 16.2B). By contrast, *Vasa* is expressed in free cells scattered throughout the zooidal hemocoel and vascular network (Fig. 16.2A). This suggests that *Vasa*-positive hemoblasts in the hemocoel are likely PGCs, although this theory should be confirmed using direct lineage tracing and/or transplantation assays.

In *P. misakiensis*, *Vasa*-expressing germline precursor cells appear *de novo* in developing zooids (Fig. 16.2A). When buds are treated with double-stranded *Vasa* RNA, zooids lack *Vasa*-positive cell aggregates and gonads, indicating that *de novo Vasa* expression is essential for gonad formation. In *B. primigenus*, germline precursor cells disappear completely from colonies when vascularization is induced by extirpating zooids and buds. However, both *Vasa* expression and germ cells can regenerate in the modified colonies within ~2 weeks. It is likely that germ cells come not only from *Vasa*-positive hemoblasts, but also from *Vasa*-negative coelomic cells in the hemocoel.

In addition, *nanos* and *Piwi* have been characterized in *B. primigenus* (Fig. 16.2A). *Nanos* is expressed widely by somatic multipotent cells, coelomic undifferentiated cells, and germ cells, especially male germ cells in the gonadal space. Consistent with this expression pattern, when siRNA is used to reduce *nanos* expression, the most severe effect observed was on spermatogenesis. *Piwi* is expressed by germ cells and coelomic undifferentiated cells. In germ cells, the *Piwi* and *Vasa* expression patterns completely overlap, while in the hemocoel, *Piwi*-expressing cells are distributed more widely than are *Vasa*-expressing cells. Reducing the levels of *Piwi* mRNA leads to defects in germline precursor cells. Therefore, in *B. primigenus Piwi/Vasa*-expressing cells also serve as germline stem cells reserved in the hemocoel.

In *B. primigenus*, gonadal tissues also express *Myc* (Fig. 16.2A). As in the case of Vasa, nanos, and Piwi, Myc is produced not only by germ cells, but also by undifferentiated somatic cells. A reduction in *Myc* transcript levels by RNA interference conspicuously lowers *Piwi* expression and results in the loss of germline precursor cells without affecting *Vasa*-positive oocyte formation. Thus, *Myc* may contribute to gonadal tissue formation via *Piwi* maintenance. Female germ cells can develop from homing *Vasa*-positive cells in the blood, and other gonadal components can arise from coelomic $Vasa^-/Piwi^+/Myc^+$ cells (Fig. 16.2A). However, the roles of these genes in gonad regeneration remain to be determined in future studies.

16.3   THE DEVELOPMENT OF COLONY SPECIFICITY

16.3.1   Fusion and Rejection Reactions Within Colonies

In 1903, Bancroft reported that *Botryllus schlosseri* colonies of different origins did not fuse together after grafting, whereas fragments from any one single colony easily fused together. Among Fl colonies developing from larvae released by one parental colony, some fused, but others did not. About 50 years later, the phenomenon of colony specificity was revisited in extensive studies by Watanabe and his colleagues, and later by Weissman and his colleagues. Colony specificity has attracted attention because it offers significant insights into the origin of the allorecognition histocompatibility (HC) of vertebrates. Recent studies appear to have identified candidate genes involved in colony specificity.

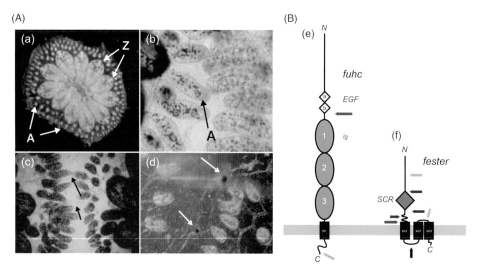

FIGURE 16.3   Colony recognition in *Botryllus schlosseri*. (A) (a) An adult *B. schlosseri* colony (left panel) consists of asexually derived individuals called zooids (Z) united by a common blood supply. (The colors are caused by pigment cells in the blood). The vasculature stops at the periphery of the colony in small protrusions called ampullae (A), where the histocompatibility reaction takes place. (b, c) When two colonies grow close together, the ampullae reach out and interact, leading to one of the two outcomes. Either the two ampullae will fuse (c, arrows), allowing the circulatory systems of the two colonies to interconnect, or else they will reject each other (d). Rejection comprises a localized inflammatory reaction where blood cells leak from the ampullae and begin killing each other, resulting in "points of rejection" that consist of dark spots formed by melanization during the reaction. Histocompatibility responses take ~24 h to occur. (Adapted from De Tomaso *et al.* (2005) *Nature*, **438**, 454–459.) (B) Topological models of the Fu/HC and fester proteins based on predictions from multiple algorithms. (e) The Fu/HC protein contains two tandem EGF domains (yellow diamonds), followed by three tandem immunoglobulin (C2 type) domains (blue ovals), a transmembrane (TM) region, and a short intracellular tail. Two splice variants of Fu/HC have been detected. The green arrow indicates a splice site that creates a putative secreted form of the protein containing about 50% of the ectodomain, while the yellow arrow points to a region where a small exon, containing no obvious motifs, is inserted in ~5% of the transcripts. (f) *fester* encodes proteins with extracellular SCR domains (green diamond), although there is no amino acid homology between the two SCR regions. This sequence is followed by three predicted TM domains and an intracellular COOH tail. Portions of the molecule deleted by alternative splicing are shown (arrows). Red and yellow arrows show regions that are frequently and rarely spliced, respectively. The green arrow on fester depicts a splice variant that removes all of the TM domains, creating a putatively secreted form of the protein. (Adapted from McKitrick and De Tomaso (2010) *Semin Immunol*, **22**, 34–38.)

When two pieces from a single colony are placed in juxtaposition at natural growing edges or artificially cut surfaces, they fuse to form a common vascular system. This phenomenon is called fusion (Fig. 16.3b, c). By contrast, a similar juxtaposition of tissues from different colonies usually results in necrosis (Fig. 16.3d). This phenomenon is termed the nonfusion reaction (NFR), or simply, rejection. Collectively, these phenomena are referred to as colony specificity, and are observed in most colonial ascidians. In *B. schlosseri*, ampullae play key roles in the NFR, together with blood cells and humoral factors.

16.3.2   A Single Polymorphic Locus *Fu*/HC

Colony fusion or rejection appears to be controlled by a single highly polymorphic locus: *Fu/HC*. Several criteria have been put forth during the search for the candidate colony specificity gene. First, the gene must be highly polymorphic. Second, the polymorphisms must correlate to fusion and rejection outcomes in both wild-type and laboratory colonies, with fusing individuals sharing at least one allele and rejecting colonies sharing no alleles. Third, the segregation of polymorphisms in the gene must correlate 100% with HC outcomes.

An extensive positional cloning effort by De Tomaso and his colleagues succeeded in isolating a candidate gene, *cFuHC*, from *B. schlosseri*. The *cFuHC* gene spans 33 kb and consists of 31 exons. The gene encodes a large (1008 amino acids) type I transmembrane protein containing a signal sequence, two EGF domains of about 750 amino acids, potentially three Ig domains followed by a transmembrane domain and an intracellular tail with no known signaling or other domains (Fig. 16.3e). Besides the full-length form of the protein, there are two alternative splice variants, sFuHC for a secreted isoform and mFuHC for a membrane-bound form. The *cFuHC* gene segregates with HC in all mapping crosses, predicts the outcome of interactions between wild-type colonies, and is expressed in tissues directly associated with the natural transplantation reaction.

As a fusion event and the formation of a chimeric colony will occur if the two colonies share one or both alleles at *Fu/HC*, in order for allorecognition specificity to be achieved, a system must be in place that can discriminate between hundreds of *Fu/HC* alleles. A Wu–Kabat analysis of the polymorphism distribution reveals that the majority of *Fu/HC* alleles are different by about 25–50 amino acids spread throughout the ectodomain, and thus it seems unlikely that recognition occurs via homotypic interactions. A candidate effector system gene, *fester*, is located over 100 kb from *Fu/HC*. The full-length *fester* gene is 11 exons long and encodes a protein of 368 amino acid residues. The fester comprises a type I transmembrane protein with a signal sequence, seven extracellular exons, an SCR or sushi domain located in extracellular exon 5, three exons (exons 8–10) that each encode a predicted transmembrane helix, and exon 11, which encodes a short intracellular tail (Fig. 16.3f). The gene is polymorphic, polygenic, and somatically diversified by alternative splicing. Similar to *Fu/HC*, *fester* expression is restricted to all tissues known to be involved in allorecognition. An siRNA-mediated decrease in expression of fester splice variants leads to a "no reaction" phenotype between histocompatible and incompatible colonies. Neither colony receives a signal initiating a response, suggesting that fester acts as an activating receptor required for the initiation of either a fusion or a rejection event. Thus, fester appears to play a dual role in the allorecognition response, functioning both as an activating receptor, required to initiate both fusion and rejection, and as an inhibitory receptor involved in self-recognition and the inhibition of a rejection response.

16.3.3   Botryllus Histocompatibility Factor

As discussed earlier, it has been disclosed that a single genetic locus, *Fu/HC*, with hundreds of codominantly expressed alleles, is responsible for fusibility/HC of *B. schlosseri* colonies. On the other hand, Rinkevich *et al.* (2012) have expressed reservations about this result. They have studied this question by cloning of genes, *in situ* hybridization, semi-quantitative PCR, and the incongruence that emerged between fusion/rejection profiles and cFu/HC-segregated polymorphism. Their results separately and cumulatively contradicted the original result. It is likely that *Botryllus* HC properties are not signaled in the claimed cFu/HC and that the cFu/HC gene is not likely to be the allodeterminant for the *Botryllus* HC locus.

A recent comprehensive sequencing of whole-transcriptomes and draft genome of genetically-defined *B. schlosseri* lines has identified a single gene that encodes self-nonself and determines graft outcomes in this organism. The gene, named "*Botryllus* histocompatibility factor" (*BHF*), is located ∼62 kb away from sFuHC and mFuHC. *BHF* is composed of three exons and encodes a highly charged and partially unstructured 252–amino acid protein, with no detectable domains or signal peptide. This gene is highly polymorphic, highly expressed in the vasculature, significantly up-regulated in colonies poised to undergo fusion and/or rejection, and is functionally linked to histocompatibility outcomes.

In the jawed vertebrates, the MHC is a haplotype, each sublocus of which specifies a different recognition process, usually by unique subsets of cells. By contrast, the *B. schlosseri* Fu/HC locus is a single gene (BHF) embedded in a haplotype of several genes with high polymorphism. BHF has none of domains that are conserved throughout protein evolution. Because BHF does not follow biological precedence by either sequence or domains, future investigations of this gene will likely reveal new mechanisms of recognition.

SELECTED REFERENCES

BANCROFT, F.W. (1903) Variation and fusion of colonies in compound ascidians. *Proc Calif Acad Sci Ser 3*, **3**, 137–186.

BROWN, F.D., SWALLA, B.J. (2012) Evolution and development of budding by stem cells: Ascidian coloniality as a case study. *Dev Biol*, **369**, 151–162.

DE TOMASO, A.W., NYHOLM, S.V., PALMERI, K.J., ISHIZUKA, K.J. *et al.* (2005) Isolation and characterization of a protochordate histocompatibility locus. *Nature*, **438**, 454–459.

KAWAMURA, K., SUGINO, Y., SUNANAGA, T., FUJIWARA, S. (2008) Multipotent epithelial cells in the process of regeneration and asexual reproduction in colonial tunicate. *Dev Growth Diff*, **50**, 1–11.

KAWAMURA, K., SUNANAGA, T. (2011) Role of Vasa, Piwi, and Myc-expressing coelomic cells in gonad regeneration of the colonial tunicate, *Botryllus primigenus. Mech Dev*, **128**, 457–470.

KAWAMURA, K., TIOZZO, S., MANNI, L., SUNANAGA, T. *et al.* (2011) Germline cell formation and gonad regeneration in solitary and colonial ascidians. *Dev Dyn*, **240**, 299–308.

KAWAMURA, K., FUJIWARA, S. (1995) Cellular and molecular characterization of transdifferentiation in the process of morphallaxis of budding tunicates. *Semin Cell Biol*, **6**, 117–126.

RINKEVICH, B., DOUEK, J., RABINOWITZ, C., PAZ, G. (2012) The candidate Fu/HC gene in *Botryllus schlosseri* (Urochordata) and ascidians' historecognition—an oxymoron? *Dev Comp Immunol*, **36**, 718–727.

MCKITRICK, T.R., DE TOMASO, A.W. (2010) Molecular mechanisms of allorecognition in a basal chordate. *Semin Immunol*, **22**, 34–38.

NAKAUCHI, M. (1982) Asexual development of ascidians: its biological significance, diversity and morphogenesis. *Am Zool*, **22**, 753–763.

SCOFIELD, V.L., SCHLUMPBERG, J.M., WEISMANN, I.L. (1982) Colony specificity in the colonical tunicate Botryllus and the origin of vertebrate immunity. *Am Zool*, **22**, 783–794.

VOSKOBOYNIK, A., NEWMAN, A.M. COREY, D.M., SAHOO, D. *et al.* (2013) Identification of a colonial Chordate histocompatibility gene. *Science*, **341**, 384–387.

WATANABE, H., TANEDA, Y. (1982) Self and non-self recognition in compound ascidians. *Am Zool*, **22**, 775–782.

CHAPTER 17

EVOLUTIONARY DEVELOPMENTAL GENOMICS

The evolutionary modifications that phenotypically distinguish groups of living organisms often reflect underlying molecular changes that affect the embryonic development of those groups. Therefore, one must first understand embryonic development in order to accurately interpret aspects of metazoan evolution. Recent advances in ascidian developmental genomics, as described in previous chapters, provide a framework within which we discuss current theories on ascidian evolution. The evolutionary aspects of ascidian development can be approached from at least three levels—with respect to the phylum (Chordata), the subphylum (Urochordata), and the species. Several topics at each level are discussed in this chapter.

17.1 THE ORIGIN AND EVOLUTION OF CHORDATES

17.1.1 The Origin of Chordates

As introduced in Chapter 1, ascidians belong to the subphylum Urochordata (Tunicata) of the phylum Chordata. The two other Chordata subphyla are Cephalochordata and Vertebrata (Fig. 1.2 and Fig. 17.1). The origin and evolution of chordates have long been debated in relation to the origin of vertebrates, the subphylum to which human beings belong. The two main alternative scenarios proposed for chordate origin include either a sessile ancestor or a free-living ancestor. The former scenario hypothesizes that ancestral deuterostomes were sedentary, like modern pterobranch hemichordates, and that chordates evolved from the motile larval stage of sessile ascidians by pedomorphosis, a form of heterochrony roughly equivalent to neoteny. On the other hand, the latter scenario invokes progressive alteration of motile adults, namely, a free-living, enteropneust hemichordate-like deuterostome ancestor evolving into a cephalochordate-like chordate ancestor.

Recent molecular phylogenetics based on the sequenced genomes and transcriptomes of echinoderms, hemichordates, cephalochordates, urochordates, and vertebrates strongly supports the notion that echinoderms and hemichordates form a clade called Ambulacraria and that, among members of another distinct clade of chordates, cephalochordates represent the most basal extant chordate lineage, with urochordates sister to vertebrates (Olfactores; Fig. 17.1). Judging from the modes of embryogenesis and adult morphologies in these groups, it is highly likely that the chordate ancestor was a free-living, acorn worm-like animal (Fig. 17.1).

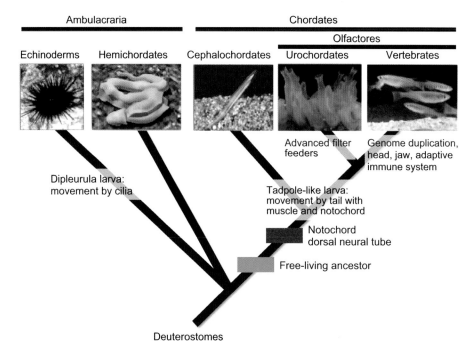

FIGURE 17.1 Schematic representation of the origin and evolution of chordates. Representative developmental events associated with the evolution of chordates are included.

17.1.2 Genomic Alteration and Chordate Evolution

A genomic comparison among the three chordate subphyla suggests that there was (i) a close evolutionary relationship between cephalochordates and vertebrates and (ii) a conspicuous evolutionary change leading to the urochordate lineage. First, synteny is well conserved between the genomes of the cephalochordate *Branchiostoma floridae* and vertebrates. The vertebrate ancestral chromosomal organization can be inferred from quadrupling the synteny blocks found in the *B. floridae* genome (Fig. 17.2B). This analysis provides direct evidence for the occurrence of a double round of genomic duplication during vertebrate evolution, although the exact timing of the duplication events is still unclear. It remains to be determined whether the duplications occurred prior to the diversification of all vertebrate groups, or after the agnatha group diverged from the coelenterates (lamprey and hagfish). Together with the overall similarity of developmental modes between cephalochordates and vertebrates, the direct evolution of vertebrates from a cephalochordate-like ancestor is easily conceived by duplicating the genome twice and developing jaws, vertebrae, an innate immune system, and other traits (Fig. 17.1). On the other hand, the *Ciona* genome shares few syntenies with vertebrate genomes.

Second, presumably reflecting the shorter generation time associated with their life style of advanced filter feeders, the size of the urochordate genome has been reduced and has become more compact. For example, representative genome sizes include 814 Mbp for the sea urchin *Strongylocentrotus purpuratus*, 520 Mbp for *B. floridae*, and 160 Mbp for *Ciona intestinalis*. This trend is remarkable in the case of larvaceans; the *Oikopleura dioica* genome is 70 Mbp and their generation time is short: ~4 days at 20 °C. In the *Oikopleura* genome, the introns are very small (with a maximum of 47 bp), as are the intergenic spaces, partly because of numerous operons. Genes outside operons are also densely packed, and the density of transposable elements is low.

In addition, there is a distinct gene loss in the urochordate genome. Comparison among *B. floridae*, *Homo sapiens*, and *C. intestinalis* genomes suggests that the last common ancestor of the chordates possessed 103 homeobox genes and that 6, 3, and 21 genes were lost during the evolution of the Olfactors, vertebrates, and tunicates, respectively (Fig. 17.2A). It may be said that *Ciona* belongs to a fast-developing animal group with a compacted genome, with loss of several TF and STM genes. On the other hand, vertebrates belong to a slow-developing animal group with expanded genomes. Further studies therefore should investigate how these genomes with their different compositions can still yield the fundamentally common chordate body plan, as seen in *Ciona* tadpoles.

17.1.3 Significance of Domain Shuffling in Evolution of Chordates

As mentioned above, the question of how novel structures are created by changing genetic information is one of the most challenging issues in evolutionary developmental genomics. It is generally accepted that the morphological features of various multicellular animals are built on a common set of regulatory genes and signaling pathways, and that the novel features emerged as a result of altered gene expression patterns. Novel genetic material has also contributed to the evolution of novel structures, and attention has been focused on gene duplications as a mechanism for the evolution of novel genetic material. Particularly in the case of vertebrate evolution, whole genome duplications that occurred twice in ancestral vertebrates have been regarded as the main force driving the evolution of novel structures. The domain shuffling of proteins is an additional mechanism to create novelties, and several different molecular mechanisms for domain shuffling have been proposed. Since the domains are often correlated with exon boundaries, exon shuffling is believed to be one of the major forces driving domain shuffling.

A comparative genomics of echinoderm, cephalochordate, urochordate, and vertebrates highlights the significance of domain shuffling as one of the mechanisms that contributes to the evolution of vertebrate- and chordate-specific characteristics (Fig. 17.3). The study identified ~1000 new domain pairs in the vertebrate lineage, including ~100 that are shared by all seven of the vertebrate species examined. Some of these pairs occur in the protein components of vertebrate-specific structures, such as cartilage and the inner ear, suggesting that domain shuffling made a marked contribution to the evolution of vertebrate-specific characteristics (Fig. 17.3e–h). The evolutionary history of the domain pairs is traceable; for example, the Xlink domain of aggrecan, one of the major components of cartilage, was originally utilized as a functional domain of a surface molecule of blood cells in chordate ancestors, and it was recruited by the protein of the matrix component of cartilage in the vertebrate ancestor (Fig. 17.3e). Some of the genes that were created as a result of domain shuffling in ancestral chordates are involved in the functions of chordate structures, such as the endostyle, Reissner's fiber of the neural tube, and the notochord (Fig. 17.3b–d).

17.1.4 Developmental Modes of Chordates in Relation to their Origin

Starting with the evolutionary scenario of a free-living ancestor suggested by genome-wide molecular phylogeny, how can we link chordate developmental processes to their evolutionary origins?

(A)

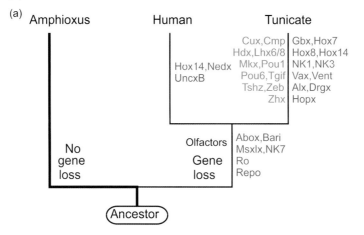

(B)

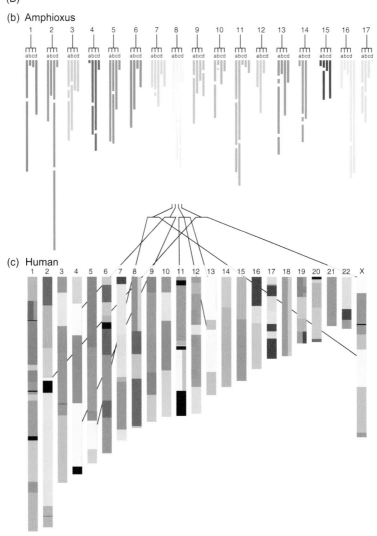

17.1.4.1 Characteristic Features of Chordates Traditionally it has been argued that chordates are characterized by possession of a notochord, a dorsal hollow neural tube or nerve cord, branchial (pharyngeal or gill) slits, an endostyle, and a postanal tail. However, recent studies reveal that the branchial slits are shared by hemichordates and echinoderms, suggesting that this feature is common to deuterostomes. The endostyle is seen only in cephalochordates, urochordates, and the larval stage of lampreys. A recent hypothesis maintains four characteristic features as synapomorphies of chordates. Specifically, these are the notochord, the dorsal hollow neural tube, the somite, and the postanal tail, although the somite in a strict sense is likely to disappear in the urochordate lineage. It should be noted that branchial slits and endostyle are organs associated with the digestive system and that all of the four synapomorphic features are associated with the evolution of tadpole-type or fish-like larvae, as discussed in following text.

17.1.4.2 The Occurrence of Fish-Like Larvae and Chordate Evolution Most marine invertebrates exhibit a biphasic life history—a larval stage and an adult stage. This biphasic lifestyle provides two opportunities for a species to invade new niches. All marine nonchordate invertebrates, irrespective of whether they are protostomes or deuterostomes, develop larvae that swim with cilia or a ciliary band (e.g., the pluteus of sea urchins and the tornaria of acorn worms). By contrast, chordate larvae are fish- or tadpole-like, and swim by tail beating instead of ciliary motion (Fig. 17.1). This alteration in the larval swimming mode represents an evolutionary developmental event that provides a clue toward understanding the origins of chordates.

Molecular support for the significance of larval form changes in chordate ancestry comes from actin gene diversification. Actins are categorized into two major groups, cytoplasmic and muscle, depending on their physiological roles. Cytoplasmic and muscle actin isoforms are encoded by different genes. In vertebrates, cytoplasmic and muscle actins are different at about 20 amino acid positions, which distinguish the two types. In nonchordate deuterostomes and protostomes, these diagnostic positions do not distinguish between muscle and cytoplasmic actins, rather, both isoforms are similar to the vertebrate cytoplasmic actins. The evolution of muscle-type actin is evident in cephalochordates and urochordates, since they have independent muscle-type and cytoplasmic-type actin genes. It is highly likely that the occurrence of fish/tadpole-type larvae during the evolution of chordates accompanied the developmental innovation of vertebrate-type muscle actin.

←───

FIGURE 17.2 (A) The loss of homeobox genes during the evolution of tunicates. Homeobox gene loss has been extensive during the evolution of the Olfactores (vertebrates plus tunicates) and tunicates since the last common ancestor of the chordates. Blue, ANTP class; red, PRD class; green, other classes. Gene losses along the tunicate branch occurred before the last common ancestor of *Ciona* and *Oikopleura*. (Adapted from Holland *et al.* (2008) *Genome Res*, **18**, 1100–1111.) (B) Quadruple conserved synteny between amphioxus (b) and vertebrate genomes (c). Partitioning of the human chromosomes into segments with defined patterns of conserved synteny (c) to amphioxus (*Branchiostome floridae*) scaffolds (b). Numbers 1–17 in (b) represent the 17 reconstructed ancestral chordate linkage groups, and letters a–d represent the four products resulting from two rounds of genome duplication. Colored bars in (c) are segments of the human genome, grouped by ancestral linkage group (above), and in the context of the human chromosomes. (Adapted from Putnam *et al.* (2008) *Nature*, **453**, 1064–1071.)

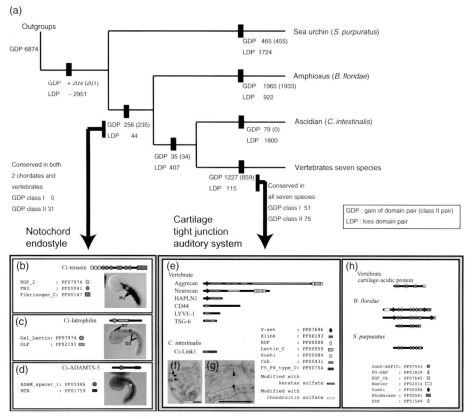

FIGURE 17.3   Schematic diagram showing significance of domain shuffling during the evolution of chordates. (a) Gain domain pair (GDP) and loss domain pair (LDP) are mapped on the deuterostome phylogenetic tree. Numbers on each node present GDP and LDP. Gains of class-II domain pairs are shown in brackets. Arrows indicate domain pairs acquired in the ancestral chordates, and those in ancestral vertebrates. Among 1227 pairs that were acquired in the ancestral vertebrates, 51 class-I domain pairs and 75 class-II domain pairs are conserved in seven vertebrate species (middle). Some of these are involved in vertebrate-specific characters such as cartilage, tight junctions, and auditory systems. Among 256 domain pairs unique to the chordates, 31 pairs are conserved in ascidian, amphioxus, and more than one species of vertebrates. Some of them are involved in chordate-specific characters such as the notochord and endostyle. (b–d) Tenascin, latrophilin, and ADAMTS-5, which were created by domain shuffling in ancestral chordates, are expressed in the notochords of ascidian embryos. Symbols are indicated with the domain ID and Pfam accession numbers. (b) *Ci-tenascin* is expressed in the ascidian notochord (N) and muscle cells (Mu). (c) *Ci-latrophilin* is expressed in the notochord, neural tube (NT), and endodermal strand (ES) of ascidian larvae. (d) *Ci-ADAMTS-5* is expressed in the ascidian notochord. (e) Domain structures of the proteins that include an Xlink domain. (f, g) *Ci-link1* is expressed in some of the blood cells (arrowheads) in an ascidian juvenile. Scale bars, 50 μm. En, Endostyle; GS, gill slits; St, stomach; It, intestine. (h) Domain structures of cartilage acidic protein and proteins containing an ASPIC-and-UnbV domain encoded by amphioxus gene models. (Adapted from Kawashima *et al.* (2009) *Genome Res*, **19**, 1393–1403.)

17.1.4.3 The Dorsal Neural Tube First, we should note the significance of *neural tube* formation in association with chordate evolution. In chordate embryos, the presumptive cells of the CNS are separated from ectodermal epidermal cells, flatten on the dorsal side to form the neural plate, and then coalesce to create a closed neural tube (see Chapter 10). This mode of dorsal hollow neural tube formation is not found in nonchordate deuterostomes and protostomes. The vertebrate brain marks the most complex neural network structure. The complexity of the vertebrate brain has been attained by numerous foldings of the neural epithelium, where cells proliferate and migrate inside to form several layers. The establishment of this complexity appears to be infeasible if the basic configuration is not a "tube." This is completely different from the insect central ganglion which does not pass through a step of tube formation. Therefore, we should consider the CNS formation as an evolutionary innovation of chordates.

The question then becomes, "how did neural tube formation occur during the evolution of deuterostomes?" The nervous system of nonchordate deuterostome larvae consists of the apical organ and ciliary bands. Because the larval apical organ degenerates at metamorphosis and new central nervous centers take over, the apical organ does not necessarily need to be discussed in relation to the chordate CNS. However, another structure, the ciliary band of dipleurula (auricularia) larva, has been a topic of discussion in relation to the origin of the chordate CNS, since Garstang (1928) proposed that during chordate evolution, the ciliary band of a dipleurula larva moved dorsally, finally fusing mid-dorsally, and enclosing the entire larval aboral surface (Fig. 17.4A). This "auricularia" hypothesis has been recently revived in light of molecular data that suggest the occurrence of a "dorso-ventral (D-V) axis inversion" between protostomes and chordates.

In vertebrates, the CNS runs dorsally, parallel to the digestive system. By contrast, the CNS in protostomes runs ventral to the digestive system. Recent findings show that BMPs and their antagonists are responsible for the establishment of D-V axes in both protostomes and vertebrates, and that the expression pattern of BMPs and their antagonists is opposite between the two groups. Namely, the D-V axis of vertebrate bodies is inverted compared to that of protostomes or nonchordate deuterostomes. Recent studies clearly show that the expression pattern of BMPs and their antagonists in amphioxus embryos is quite similar to that seen in amphibian embryos, but that the corresponding expression patterns in hemichordate embryos are inverted. Therefore, based on the BMP/antagonist expression profiles, the D-V axis inversion appears to occur during chordate evolution. No consensus views however have emerged to explain the mechanism(s) underlying the observed inversion of BMP/antagonist expression during the evolution of chordates.

However, the inversion of the D-V axis during chordate evolution cannot necessarily explain the appearance of the neural tube on the dorsal side of the ancestral chordate embryo (Fig. 17.4A). A revised, though somewhat unlikely, hypothesis based upon dipleurula larvae suggests that the chordate CNS evolved from the postoral loop of the ciliary band. On the other hand, the recently proposed "aboral-dorsalization" hypothesis emphasizes that the morphogenesis required to form the dorsal neural tube in chordate embryos is completely different from that required to form the ciliary band nervous systems of nonchordate deuterostome embryos (Fig. 17.4A). This is particularly evident during neural tube formation in amphioxus embryos. The aboral-dorsalization hypothesis proposes that these structures were formed *de novo* during chordate evolution, and that the formation of these structures on the dorsal side of the embryo was the result of necessity due to limited space on the opposite (oral) side (Fig. 17.4d–f). Therefore, the

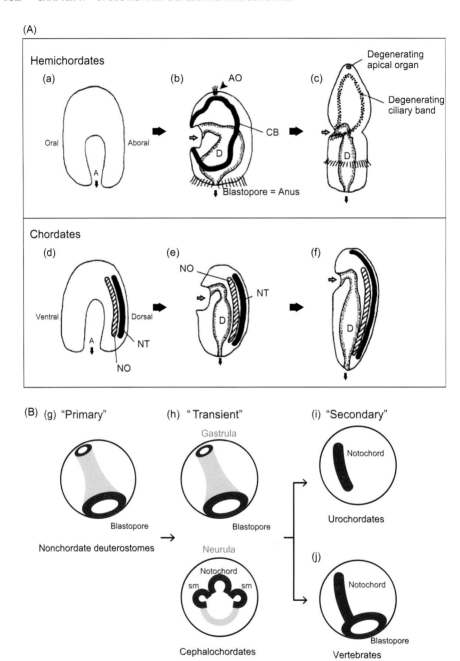

FIGURE 17.4 (A) Schematic outlining the aboral-dorsalization hypothesis explaining the development of tadpole-type larvae with a dorsal neural tube and notochord during chordate evolution. Embryogenesis of (a–c) hemichordates and (d–f) chordates. It should be emphasized that the neural tube and notochord formed on the dorsal side of the embryo are completely novel structures without any relationship to the apical organ and stomochord of hemichordate enteropneust embryos and juveniles, respectively. A, archenteron; AO, apical sensory organ; CB, ciliary band; D, digestive track; NT, neural tube; NO, notochord. White arrows indicate mouth and black arrows anus. See the main text for details. (B) The role of *Brachyury* in gastrulation and notochord formation. (g–j) The transition from primary *Brachyury* expression in the blastopore (a) to secondary expression in the notochord (i, j).

D-V axis inversion hypothesis should be further examined in relation to the *de novo* formation of the neural tube and notochord on the dorsal side of the chordate ancestral embryo.

17.1.4.4 Notochord

The fish/tadpole-type larva swims by beating the tail side to side via contractions of striated muscle. This type of larva swims faster, and thus catches prey more efficiently than do the nonchordate deuterostome larvae that use ciliary movements. As mentioned above, the molecules associated with muscle movement, such as actins, gained functional efficiency through the modifications leading to the so-called vertebrate muscle-type. The notochord also supports the beating tail, providing a stiffness that is required for efficient tail muscle function. The aforementioned synapomorphies of chordates (notochord, dorsal neural tube, somite, and postanal tail) developed along with efficient larval motility.

Interestingly, the developmental mode and structural components of the notochord differ between cephalochordates and Olfactores (urochordates + vertebrates). The cephalochordate notochord is formed by "pouching-off" from the dorsal region of the archenteron (Fig. 17.4h). In addition, the cephalochordate notochord displays muscle properties, and its constituent cells include myofibrils. An EST analysis indicates that ~11% of the genes that are expressed in the *Branchiostoma* adult notochord encode muscle component proteins, including actin, tropomyosin, troponin I, and creatine kinase. By contrast, in Olfactores the notochord is formed by convergence, intercalation, and extension of precursor cells that are bilaterally positioned in the early embryo (Fig. 17.4i, j); these cells do not possess any muscle properties (see Chapter 9). Vacuolation within the cells provides both stiffness and an increased cell volume; this is the case in both ascidians and vertebrates (Fig. 9.5). It is worth mentioning that the convergence, intercalation, and extension of notochord cells are among the most significant morphogenetic processes involved in forming the dorsal midline structures that characterize urochordates and vertebrates.

As discussed in Chapter 9, a member of the T-box family, *Brachyury*, plays an essential role in notochord formation in chordate embryos. A comparison of the genome of a unicellular choanoflagellate with the genomes of sponge and cnidarians implies that *Brachyury* appeared during the evolution of multicellular animals or metazoans. Gastrulation is a morphogenetic movement that is essential for the formation of two- or three-germ-layered embryos. *Brachyury* is transiently expressed in the blastopore region, where it confers on cells the ability to undergo invagination. This process is involved in the formation of the archenteron in all metazoans. This is a "primary" function of *Brachyury* (Fig. 17.4B). During the evolution of chordates, *Brachyury* gained an additional expression domain at the dorsal midline region of the blastopore (Fig. 17.4B). In this new expression domain, *Brachyury* attained its "secondary" function, by which recruiting another set of target genes to form a dorsal axial organ, the notochord (Fig. 17.4B). In contrast to the transient *Brachyury* expression in cells of the blastopore region, the secondary *Brachyury* expression is constitutive, and thus was able to recruit genes that became involved in notochord formation. The Wnt/β-catenin, BMP/Nodal, and FGF signaling pathways are involved in the transcriptional activation of *Brachyury*. The molecular mechanisms of *Brachyury*'s secondary function, in the context of the dorsal–ventral inversion theory and the aboral-dorsalization hypothesis, may be one of the most interesting and central evolutionary developmental biology problems in the history of metazoan evolution.

17.1.4.5 Neural Crest The neural crest is an embryonic cell population unique to vertebrates. The cells emerge from the neural plate border, migrate extensively and give rise to diverse cell lineages including melanocytes, craniofacial cartilage and bone, peripheral and enteric neurons and glia, and smooth muscle. The evolution of the neural crest has been proposed to be a key event leading to the appearance of new cell types that fostered the transition from filter feeding to active predation in ancestral vertebrates. The gene regulatory network underlying the neural crest formation appears to be highly conserved as a vertebrate innovation (Fig. 17.5A). Namely, border induction signals from the ventral ectoderm and underlying mesendoderm pattern the dorsal ectoderm, inducing expression of neural border specifiers. These inductive signals then work with neural border specifiers to upregulate the expression of neural crest specifiers. The neural crest specifiers cross-regulate and activate various effector genes, each of which mediates a different aspect of the neural crest phenotype, including cartilage, pigment cells, and peripheral neurons (Fig. 17.5A). It has been shown that amphioxus lacks most neural crest specifiers and the effector subcircuit controlling neural crest delamination and migration (Fig. 17.5A).

The presence of a true neural crest remains controversial in urochordates. In a mangrove tunicate, a migratory cell population originating from the vicinity of the neural tube and expressing neural crest markers was reported to resemble the neural crest cells, although subsequent studies on other tunicate species suggested that these migratory cells may arise from the mesoderm flanking the neural tube. It was then suggested that a mesoderm-derived mesenchyme lineage (A7.6) in *Ciona* possessed some of the properties of the neural crest. However, these cells do not arise from the neural plate border and lack expression of some of the key neural crest regulatory genes.

A recent study examined this question with sophisticated techniques of cell tracing and gene function. It became clear that *C. intestinalis* possesses a cephalic melanocyte lineage (a9.49) exhibiting a gene cascade similar to neural crest (Fig. 17.5B), in which Wnt signaling activates *FoxD* in the presumptive ocellus, and then *FoxD* represses *Mitf* independent of its DNA-binding domain, consistent with its mode of regulation in avian embryos (Fig. 17.5d, e). In addition, although *Twist* (*twist-like 2*) is not expressed in the a9.49 lineage, misexpression of this gene in the a9.49 lineage can reprogram it into migrating "ectomesenchyme" (Fig. 17.5f, g). Therefore, it is likely that the neural crest melanocyte regulatory network predated the divergence of tunicates and vertebrates. The cooption of mesenchyme determinants, such as *Twist*, into the neural plate ectoderm was crucial to the emergence of the vertebrate neural crest.

17.1.4.6 Endostyle The endostyle is a special organ of the pharynx in urochordates (Fig. 1.1b, o), cephalochordates, and larval lampreys (ammocoetes), and secretes mucoproteins into the alimentary canal for filter feeding. In addition, as the endostyle uptakes iodine, an evolutionary link has been argued between this organ and the follicular thyroid of vertebrate. This hypothesis is supported by the expression of *TTF-1* in the endostyle of *Ciona*, amphioxus, and lamprey, an ancestral transcription factor gene which controls the survival of thyroid follicular cells at the beginning of organogenesis and regulates the expression of thyroid-specific genes in vertebrate adults. Although the true structure of the thyroid appears for the first time in lamprey adults, some of the epithelial cells persist and transform into the follicles during metamorphosis of larval lampreys.

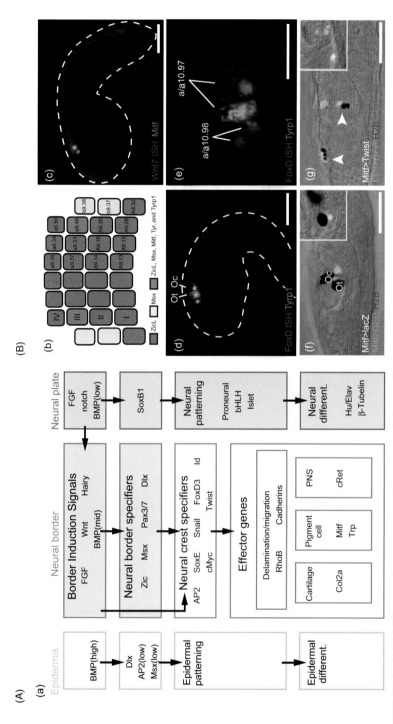

FIGURE 17.5 (A) (a) The neural crest gene regulatory network in vertebrates. Black arrows indicate empirically verified regulatory interactions. Shaded areas represent the conserved subcircuits of the respective gene regulatory networks between vertebrates and cephalochordates; namely. (Adapted from Yu (2010) *Zoology*, **113**, 1–9.) (B) (b) The neural plate of *Ciona intestinalis*, indicating the lineage-specific expression of enhancers especially in a9.49 (green). (c) Tailbud embryo electroporated with Mitf(promoter):lacZ, detected with an antibody (green) and hybridized with a Wnt7 probe (red), showing Mitf expression in melanocytes of the embryo. (d, e) Tailbud embryo electroporated with Tyrp1:lacZ detected with an antibody (green) marking the precursors of the otolith (Ot) and ocellus (Oc), and hybridized with a FoxD probe (red). *FoxD* is expressed in the posterior a10.97 cell. (f, g) Larvae electroporated with Mitf:GFP and Mitf:H2B mCherry. Insets show lineage marked with Tyr:mCherry and Tyr:H2B YFP (yellow fluorescent protein); both the Tyr and Mitf reporters label the same a9.49 cells. (f) Coelectroporated with Mitf:lacZ. (g) and Mitf:Twist. Arrowheads indicate ectopic position of a9.49 derivatives. (Adapted from Abitua *et al.* (2012) *Nature*, **492**, 104–107.)

The *Ciona* endostyle consists of nine zones, including supporting zones (zones 1, 3, and 5), three-pairs of glandular zones (zones 2, 4, and 6), and iodine-binding zones (zones 7, 8, and 9) (Fig. 17.6A), while the amphioxus one contains seven functional compartments such as supporting (zones 1 and 3), two pairs of glandular (zones 2 and 4), and iodine-binding zones (zones 5a, 5b, and 6) (Fig. 17.6B). In *Ciona*, the adult endostyle is derived from larval endoderm (Fig. 7.1c, d). Recent studies demonstrate that not only structural genes but also TF genes are specifically expressed in each zone, suggesting an involvement of these genes in the regionalization of the endostyle. For example, *TTF-1* is expressed in zones 1, 3, and 5; *TPO* in zone 7; *Pax2/5/8* in zones 3, 5, and 7; *FoxE4* in zones 5 and 7; *FoxQ1* in zones 3, 5, 7, and 8; and *FoxA* in zones 3 and 8 (Fig. 17.6b). The amphioxus endostyle shows a similar TF gene expression profile along the basal–proximal axis of the *Ciona* pharynx (Fig. 17.6d), suggests that genetic basis of the endostyle function was already in place before separation of the two lineages. In addition, TF gene expression associated with the anterior–posterior (AP) regionalization of the endostyle has been revealed in the larvacean (*O. dioica*), namely, *Otx* expresses rostrally, *Hox1* caudally, and two *Pax2/5/8* paralogs centrally (Fig 17.6C). Because the ordered expression of *Otx*, *Pax2/5/8*, and *Hox1* displays patterning in both the endodermally derived endostyle and the ectodermally derived CNS, this gene set is highly likely to a developmental genetic toolkit of stem bilaterians, which repeatedly provided AP positional information in various developmental events.

These comparative analyses suggest that the *Ciona* endostyle provides an experimental system to understand molecular mechanisms involved in the formation and function of this organ in the context of ontogenic and phylogenic developmental biology.

17.2 EVOLUTION OF UROCHORDATES

The urochordates consists of three highly divergent major groups—the ascidians, the larvaceans, and the thaliaceans. Recent molecular phylogeny evidence suggests that the thaliaceans are included in a clade of the class Enterogona (Fig. 1.2). The phylogenetic position of Appendicularia (Larvaceans) is still enigmatic; some insist that its basic position among urochordates while some place it within a clade of Pleurogona (Fig. 1.2). Although understanding the molecular embryology of thaliaceans is essential for an evolutionary discussion of urochordates, very little information on thaliacean molecular embryology exists, limiting the discussion of urochordate evolution.

17.2.1 Cellulose Biosynthesis and Tunic Formation

The early Cambrian fossil tunicate *Shankouclava* from South China exhibits a similar outer morphology to extant ascidians, suggesting that the ascidian that appeared on the earth around 520 million years ago had a tunic. As discussed in Chapter 8, bacterial gene(s) for cellulose synthase (i.e., *CesA*) horizontally transferred into the ancestral tunicate genome. The *Ciona* genome retains the *CesA* gene as a single copy while *Oikopleura* duplicated the gene in the genome. Cellulose is synthesized by a large multimeric protein complex that tracks along cortical microtubules at the plasma membrane. The only known components of these complexes are CesA proteins. Therefore, it is of urgent interest to elucidate the other protein components in the multimeric complex. It is interesting to consider whether the genes encoding the other components of the protein complex were

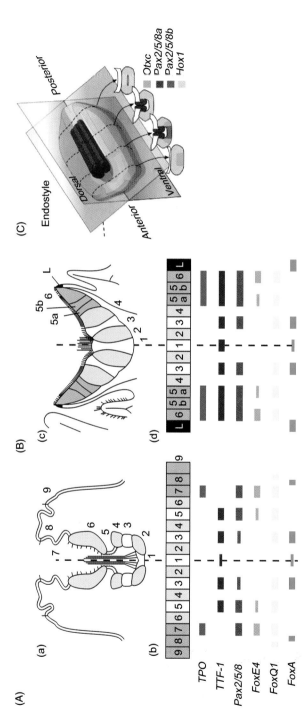

FIGURE 17.6 (A, B) Comparison of the histological components and gene expression patterns in the endostyles of *Ciona* (a, b) and the amphioxus (c, d). Transverse sections. (a, c) Both endostyles possess supporting elements, glandular elements, and thyroid-equivalent elements. Supporting elements, glandular elements, thyroid-equivalent elements, and lateral region next to zone 6 of the amphioxus endostyle are indicated in white, lightly shaded, darkly shaded, and black boxes, respectively. The midlines of the endostyle are indicated by the dotted lines. (b, d) Gene expression pattern. Each component is aligned with an equal interval and is indicated by numbers for each zone. Distinct expression domain (bold bar) and weak and/or occasional expression domain (thin bar) are shown below the corresponding zones. (Adapted from Hiruta *et al.* (2005) *Dev Dyn*, **233**, 1031–1037.) (C) Schematic three-dimensional reconstruction of the larvacean endostyle summarizing the anterior-posterior (AP) regionalization revealed by the expression domains of *Otxc* (green), *Pax2/5/8a* (red), *Pax2/5/8b* (blue), and *Hox1* (yellow), and showing virtual representations of cross-sections at different levels of the AP axis. (Adapted from Cañestro *et al.* (2008) *Dev Dyn*, **237**, 1490–1499.)

187

also horizontally transferred from bacteria or whether they originated within the ancestral tunicate genome.

What benefits do tunicates receive by their ability to synthesize cellulose? Since cellulose or tunicin is a component of the tunic, this question is similar to asking "what benefits do tunicates receive by being enclosed in a tunic?" Some ascidians, such as *Halocynthia* and *Styela*, have a hard tunic, which may function as an outer protective structure, like a mollusk shell. On the other hand, the larvacean tunic is very soft and the "house" is frequently discarded. The house functions to generate water currents to capture tiny food particles within the seawater. Therefore, a part of the tunic is not always a protective apparatus.

A recurrent evolutionary question raised here again is whether the appendicularians (larvaceans) are ancestral or divergent among urochordates. If larvaceans are ancestral tunicates and their tunic is comparatively soft and not tough enough to be protective, the original function of the tunic may be different from that observed in ascidians. One suggestion is that the tunic functioned as a fin for swimming tadpole larvae, an additional device that facilitates faster movement by tail beating, analogous to the fins of fish. In that context, *swimming juvenile* (*sj*) may give insight into another interesting evolutionary event. Namely, this mutant has juvenile organs with a larval tail. In other words, the process of *sj* formation partially resembles that of larvaceans, which maintain the basic larval form for their entire life. *sj* recalls a kind of atavism or reversion from the sessile ascidian style to a free-swimming larvacean style. Many colonial ascidians also develop juvenile organs in the trunk while they are swimming. These related phenomena should be studied to determine the underlying molecular mechanisms involved.

17.3 EVOLUTION AMONG ASCIDIANS

Ascidians, as a group, feature several developmental differences that should be discussed in an evolutionary context.

17.3.1 Solitary and Colonial Lifestyles

Molecular phylogeny based on 18 rDNA sequences clearly indicates that colonial ascidians evolved independently several times within the class. For example, among the suborder Stolidobranchia, *Halocynthia* is a solitary species while *Botryllus* is a colonial species (Fig. 1.2). Similarly, among the suborder Phlebobranchia, *Ciona* is solitary while *Perophora* is colonial (Fig. 1.2). Colonial ascidians gained the ability to reproduce asexually by budding to form colonies. It is interesting to note that the oozooid (zooid developed from an egg) reproduces only asexually and does not attain sexual maturity (as described in Chapter 16). By contrast, the blastozooid (zooid developed from a bud) reproduces both sexually and asexually. That is, without at least one phase of asexual development, colonial ascidians cannot propagate sexually, suggesting that asexual reproduction of colonial ascidians is an adaptation for propagation. This raises the intriguing question as to which gene regulatory network(s) is responsible for the state transition from oozooids to blastozooids.

Furthermore, the evolution of a colonial lifestyle is accompanied by the development of "fusion and rejection" responses when two colonies meet (as described in Chapter 16).

	Urodele species	Anural species
	Molgula oculata *Ciona intestinalis*	*Molgula occulta* *Molgula tectiformis*

Muscle

Regulatory genes

Macho-1	+	+
Tbx6	+	+

Structural (embryonic muscle) genes

Actin	+	−
MHC	+	−
MLC	+	−
Tropomyosin	+	−
TroponinT	+	−
TroponinC	+	−

Notochord

Regulatory genes

Brachyury	+	+

Structural genes

Prickle	+	+
Noto2	+	+
cdc45	+	+
tensin	+	+
leprecan	+	+

FIGURE 17.7 Changes in gene expression profiles during the evolution of anural (tailless) *Molgula* species (*M. occulta* and *M. tectiformis*) compared with urodele species (*M. oculata* and *Ciona intestinalis*).

Colonial ascidians also develop and/or retain stem-like cells with greater developmental potential as compared to those found in solitary ascidians. These phenomena raise challenging questions to be tackled by future developmental genomics studies.

17.3.2 Anural Ascidians

Several species in the families Molgulidae and Styelidae do not develop a conventional tadpole larva with a tail and larval sensory organs: instead they either form an immotile tailless larva or directly develop a juvenile body shortly after hatching (Fig. 17.7). Accordingly, they are called anural ascidians or tailless species; such organisms do not form a functional notochord or develop larval muscle during embryogenesis. Molecular phylogeny shows that anural ascidians evolved independently several times from urodele ancestors. An interesting finding is that a hybrid larva obtained by cross-fertilization of eggs of the anural *Molgula occulta* with sperm of its sister urodele species *Molgula oculata*, often develops a short tail with a notochord, suggesting that the anural mode of development evolved through relatively simple genetic changes.

Reflecting the loss of larval muscle cell formation, the expression levels of many muscle structural genes, including actin and myosin heavy chain, are below the limits of detection in anural embryos (Fig. 17.7). Furthermore, the muscle actin genes have become pseudogenes, although the promoter module of the *M. occulta* muscle actin gene is likely to retain activity since the promoter is able to drive muscle-specific expression when a reporter construct is introduced into *M. occulta* eggs. By contrast, *macho-1* mRNA is localized similarly in the eggs of the anural ascidian, *Molgula tectiformis*, and urodele ascidians. The expression of the *macho-1* downstream target, *Tbx6*, is also evident in muscle lineage blastomeres (Fig. 17.7). These results indicate that the failure of muscle cell differentiation is not due to the loss of function of the upstream genetic cascade, but rather is an effect of the gradual degradation of downstream structural genes.

Brachyury is also expressed similarly in the early embryos of *M. tectiformis*, *M. occulta*, and urodele ascidians. Fifteen of 35 orthologs of the downstream components of *Ciona Brachyury* are expressed in notochord lineage blastomeres (Fig. 17.7). This suggests that the penetration of evolutionary changes responsible for the failure of notochord formation is slower than those leading to muscle loss, even though the differentiation of the notochord is also suppressed in these species (Fig. 17.7).

Subtractive screening of transcripts expressed in the gonads of *M. oculata* and *M. occulta* identified a novel transcription factor gene, *Manx*, which is expressed in *M. oculata* gonads but sparsely, if at all, in *M. occulta* gonads. Functional suppression of *Manx* in the hybrid larva results in the failure of tail and sensory organ formation. *Bobcat*, which encodes a DEAD-box RNA helicase, is another gene identified in the subtractive screen that is also involved in the hybrid larval tail development. In addition, EST analysis shows that genes encoding proteins with no similarity to known proteins of other organisms are abundantly expressed in *M. tectiformis* embryos. Recent advances in developmental genomics technologies may allow for comparative genomic studies between *M. oculata* and *M. occulta*. It is anticipated that such studies would identify genomic changes governing their differing modes of embryogenesis.

17.3.3 Differences in the Developmental Genetic Cascades Between *Ciona* and *Halocynthia*

As discussed in previous chapters, recent studies of the molecular mechanisms of ascidian embryogenesis have shown that the regulatory genes governing developmental genetic cascades and gene regulatory networks are not always identical between *C. intestinalis* and *Halocynthia roretzi*. Notably, the embryonic cells that give rise to the so-called "secondary" muscle and "secondary" notochord employ different regulatory genes. For example, A-line muscle cell development of *Ciona* embryos relies on an FGF → Nodal → Delta/Notch signaling cascade. By contrast, in *Halocynthia* embryos, A-line muscle cell development requires FGF signaling but not Nodal or Delta/Notch.

Both secondary muscle and notochord occupy the posterior-most region of the tail. This positioning inspires speculation as to how the posterior tail region evolved—gradually by adding components step by step or in a single shift by a threshold-like genetic change, for example, a mutation in a homeotic gene. Further analyses of the developmental genomics of ascidian embryogenesis will provide insights into the evolutionary mechanisms that drove the phenotypic changes evident today.

17.3.4 Comparative Genomics of Ascidians

As mentioned earlier, decoded genomes of *C. intestinalis* and *Ciona savignyi* provide us a tool to find cis-regulatory modules as conserved noncoding sequences responsible for specific expression of developmentally relevant genes (Chapter 3). Genome projects are now underway in various species including *H. roretzi* and *Phallusia mammillata*, which may facilitate our understanding of developmental genomics of ascidian. Comparative genomic studies between *M. oculata* and *M. occulta* would identify genomic changes governing their differing modes of embryogenesis. There are two cryptic species of *C. intestinalis*, one (Sp. A) distributing world-widely (comparatively worm seawater) and the other (Sp. B) mainly from North Atlantic (comparatively cold seawater). Comparative genomics of the two cryptic species would disclose molecular mechanisms involved in the invasion of marine organisms to adapt for new niches of their life.

In conclusion, tunicates provide us various questions onto the origin and evolution of chordates. Nowadays it is possible to discuss these questions based on molecular evidence. *Ciona* and other tunicates should be further investigated by evolutionary developmental biology, taking advantages of their genomes and modes of embryogenesis.

SELECTED REFERENCES

ABITUA, P.B., WAGNER, E., NAVARRETE, I.A., LEVINE, M. (2012) Identification of a rudimentary neural crest in a non-vertebrate chordate. *Nature*, **492**, 104–107.

CARROLL, S.B., GRENIER, J.K., WEATHERBEE, S.D. (2005) *From DNA to Diversity*, Blackwell, Malden, MA.

CAMERON, C.B., GAREY, J.R., SWALLA, B.J. (2000) Evolution of the chordate body plan: new insights from phylogenetic analyses of deuterostome phyla. *Proc Natl Acad Sci USA*, **97**, 4469–4474.

CAÑESTRO, C., BASSHAM, S., POSTLETHWAIT, J.H. (2008) Evolution of the thyroid: anterior–posterior regionalization of the *Oikopleura* endostyle revealed by *Otx*, *Pax2/5/8*, and *Hox1* expression. *Dev Dyn*, **237**, 1490–1499.

DAVIDSON, E.H. (2006) *The Regulatory Genome. Gene Regulatory Networks in Development and Evolution*, Academic Press, San Diego, CA.

DELSUC, F., BRINKMANN, H., CHOURROUT, D., PHILIPPE, H. (2006) Tunicates and not cephalochordates are the closest living relatives of vertebrates. *Nature*, **439**, 965–968.

GANS, C., NORTHCUTT, R.G. (1983) Neural crest and the origin of vertebrates: a new head. *Science*, **220**, 268–273.

GARSTANG, W. (1928) The morphology of the Tunicata, and its bearings on the phylogeny of the Chordata. *Q J Microsc Sci*, **72**, 51–187.

GEE, H. (1996) *Before the Backbone. Views on the Origin of the Vertebrates*, Chapman and Hall, London.

GERHART, J., LOWE, C., KIRSCHNER, M. (2005) Hemichordates and the origin of chordates. *Curr Opin Genet Dev*, **15**, 461–467.

GYOJA, F., SATOU, Y., SHIN-I, T., KOHARA, Y. *et al.* (2007) Analysis of large scale expression sequenced tags (ESTs) from the anural ascidian, *Molgula tectiformis*. *Dev Biol*, **307**, 460–482.

HIRUTA, J., MAZET, F., YASUI, K., ZHANG, P. *et al.* (2005) Comparative expression analysis of transcription factor genes in the endostyle of invertebrate chordates. *Dev Dyn*, **233**, 1031–1037.

HOLLAND, L., ALBALAT, R., AZUMI, K., BENITO-GUTIÉRREZ, E. *et al.* (2008) The amphioxus genome illuminates vertebrate origins and cephalochordate biology. *Genome Res*, **18** 1100–1111.

JEFFERY, W.R., SWALLA, B.J. (1990) Anural development in ascidians: evolutionary modification and elimination of the tadpole larva. *Semin Dev Biol*, **1**, 253–261.

JEFFERY, W.R., STRICKLER, A.G., YAMAMOTO, Y. (2004) Migratory neural crest-like cells form body pigmentation in a urochordate embryo. *Nature*, **431**, 696–699.

KAWASHIMA, T., KAWASHIMA, S., TANAKA, C., MURAI, M. *et al.* (2009) Domain shuffling and the evolution of vertebrates. *Genome Res*, **19**, 1393–1403.

KUSAKABE, T., MAKABE, K.W., SATOH, N. (1992) Tunicate muscle actin genes: structure and organization as a gene cluster. *J Mol Biol*, **227**, 955–960.

LACALLI, T.C. (2005) Protochordate body plan and the evolutionary role of larvae: old controversies resolved?. *Can J Zool*, **83**, 216–224.

NIELSEN, C. (1999) Origin of the chordate central nervous system and the origin of chordates. *Dev Genes Evol*, **209**, 198–205.

NIELSEN, C. (2012) *Animal Evolution*, 3rd edn, Oxford University Press, Oxford.

PUTNAM, N.H., BUTTS, T., FERRIER, D.E., FURLONG, R.F. *et al.* (2008) The amphioxus genome and the evolution of the chordate karyotype. *Nature*, **453**, 1064–1071.

SATO, A., SATOH, N., BISHOP, J.D.D. (2012) Field identification of 'types' A and B of the ascidian *Ciona intestinalis* in a region of sympatry. *Mar Biol*, **159**, 1611–1619.

SATOH, N. (2008) An aboral-dorsalization hypothesis for chordate origin. *Genesis*, **46**, 614–622.

SATOH, N., TAGAWA, K., TAKAHASHI, H. (2012) How was the notochord born?. *Evol Dev*, **14**, 56–75.

SUZUKI, M.M., SATOH, N. (2000) Genes expressed in the amphioxus notochord revealed by EST analysis. *Dev Biol*, **224**, 168–177.

SWALLA, B.J. (2006) Building divergent body plans with similar genetic pathways. *Heredity*, **97**, 235–243.

YU, J.-K.S. (2010) The evolutionary origin of the vertebrate neural crest and its developmental gene regulatory network–insights from amphioxus. *Zoology*, **113**, 1–9.

INDEX

aboral-dorsalization hypothesis, 181

acraniates, 2

actin, 57, 85
  cytoplasmic-type actin, 179
  muscle-type actin, 179

actomyosin, 85, 86

adaptive immune system, 2, 159

adhesive organ (papillae), 72, 73, 99, 100,

adult gene cluster, 41

adult morphology, 1, 17
  body-wall muscle, 137
  heart muscle, 137
  nervous system, 101–104
  siphon muscle, 143

AFF, 83

agranular hemocytes, 162

aimless, 36, 84

AKR1a, 109

A-line cells, 16

a-line cells, 15

allelic polymorphism, 19

allorecognition, 159

allorecognition histocompatibility (HC), 171

α-tropomyosin 1 and 2, 58

alternative splicing, 172

Ambulacraria, 175

amphioxus, 2, 160, 176

ampullae, 155, 173

animal–vegetal (An–Vg) axis, 9, 44, 113

ANISEED, 27, 118, 126

antagonistic action, 115

anterior larval tail muscle cells (ATMs), 138

anterior–posterior (A–P) axis, 9, 44, 113

antidorsalizing morphogenetic protein
  (ADMP), 71, 72, 101, 127

antisense morpholino oligonucleotides
  (MOs), 31–33

anural ascidians, 189

anus, 1

AP4, 84

aPKC, 44

Aplousobranchia, 4

Appendicularia (larvacean), 3, 188

Arabidopsis thaliana, 132, 133

Aristotle, 4

Ascidiacea, 3

Ascidia gemmata, 162

ascidians, 1–7

Ascidia sydneiensis samea, 164

Ascidiella aspersa 5, 6

asexual reproduction, 167–169

asymmetric divisions, 50, 140, 146

atrial siphon muscles (ASMs), 143

atrial siphon placode, 129

auricularia hypothesis, 181

autonomous specification (differentiation),
  53, 65

axial organ, 77

axoneme, 148

axons extension, 84, 95

bacterial artificial chromosome (BAC)
  clones, 23, 24

basic helix-loop-helix (bHLH) family
  proteins, 22

basic leucine zipper domain (bZIP) family
  proteins, 22

β-catenin, 33, 64, 81, 115, 126, 127, 147

bilateral symmetry, 10

biphasic lifestyle, 179

blastozooid, 167

B-line cells, 16

b-line cells, 16
  B7.5-line cells, 138
  B7.6-line cells, 145

blood cell, 18, 107, 159, 162–165

Bmp2/4, 95, 141

Bobcat, 190

body plans, 113, 119

body-wall muscle, 107, 137

Developmental Genomics of Ascidians, First Edition. Noriyuki Satoh.
© 2014 John Wiley & Sons, Inc. Published 2014 by John Wiley & Sons, Inc.

bone morphogenetic protein (BMP), 71, 121, 181
  BMP signaling, 95, 121, 183
Botrylloides, 167
Botrylloides violaceus, 169
Botryllus, 167, 188
Botryllus primigenus, 169–171
Botryllus schlosseri, 4, 171–174
Brachyury, 38, 79–83, 121, 126, 127, 134, 183
brain, 89, 99
branchial basket (sac), 1, 18, 63
branchial (pharyngeal or gill) slits, 179
Branchiostoma floridae, 176–179, 183
bud (budding), 167

C3, 160
C6-like, 160
Ca^{2+} burst, 152
Caenorhabditis elegans, 21, 31, 46, 49, 147
cardiogenesis, 141
Cdc25, 95
Cdc42, 83, 141
Cdx, 95
cell-autonomous cascade, 79
cell–cell communication, 119
cell–cell signaling pathway molecules (SPMs), 7
cell cycle, 95
cell division-required event, 158
cellulose, 1, 73–75, 186, 188
  biosynthesis, 73–75
  CesA, 73–75, 186, 188
central and peripheral nervous systems, 4, 89
central nervous system (CNS), 4, 11, 89–100
centrosome-attracting body (CAB), 44, 46, 50, 115
Cephalochordata, 2, 117, 175, 184
cephalic melanocyte, 184
cerebral ganglion, 1, 18, 89, 102
Chabry, Laurent, 5
chemotaxis, 28, 152
Chenjan Fauna, 2
choline acetyltransferase (ChAT), 97, 98
chongmague, 36, 84
Chordata, 2, 77, 175–180
chorion, 9, 150
chromatin immunoprecipitation (ChIP), 126
chromosomal level regulation, 126
chromosomal map, 23–26
ciboulot (thymosin beta), 84
ciliary band, 179

Ciona intestinalis, 2, 4, 5, 7, 9, 13, 16, 17, 19–29, 31, 33–38, 41, 42, 45, 46, 50, 55, 57–61, 65, 70–73, 79, 80, 82, 85, 86, 89, 92, 94, 97, 98, 100–110, 113, 118, 121–130, 132–138, 141–149, 150–161, 163, 176, 177, 179, 184–186, 190, 191
  Ciona intestinalis Gene Collection Release 1 (CiGCR1), 26
Ciona savignyi, 22, 34, 35, 36, 46, 50, 55, 57, 58, 73, 78, 84, 97, 152, 190
CIPRO database, 27
Clavelina, 169
cleavage, 10
cleavage-arrested embryo, 5, 53
clock mechanisms, 131
cnidarians, 64
COE, 127, 143
coelenterates, 176
coelomic cells, 18
coin-shaped cell, 77
collagen II, 83
collagen XI, 83
collagen XVIII, 83
collagen XXII, 83
colinearity, 129
colonial (compound) ascidians, 4, 167–174, 188
colony specificity, 171–174
comparative genomics, 190
compartment cell, 162
competence, 121, 155
complement components, 159–161
Conklin, Edwin G, 5, 10, 53
convergence (notochord), 79, 84
CpG islands, 131
creatine kinase, 58
cryptic species, 4, 191
CTD-Ser2, 147, 148
C-type lectin, 162
cyclic AMP response-element (CRE), 57
Cynthia (Styela) partita, 53
Cyp26, 94
cytoplasmic-type actin, 179

Darwin, Charles, 4
Delta (Notch/Delta), 59, 82, 190
Delta-like, 82, 126
Delta2, 59, 61, 71, 91, 92, 118
DiI, 32
dipleurula larva, 181
Distaplia, 169
Dlx-c, 84
Dmbx, 95

DNA methylation, 131–133
DNA replication, 131
domain shuffling, 177
dorsal (hollow) neural tube, 2, 179, 181
dorsal strand plexus, 102
dorsal tail region, 99
dorsal–ventral (D–V) axis, 9, 44, 113
  D-V axis inversion, 181
Driesch, Hans, 5
Drosophila hydei, 36
Drosophila melanogaster, 5, 21, 31, 41, 137,
    145, 147
Dsh, 83
dynein, 148

E-box, 57
ectopic gene expression, 32
effector system, 172
egg fragmentation, 55
electroporation, 33, 34
elongation (of notochord), 85
embryonic and adult gene cluster, 41
embryonic axis, 9
embryonic gene cluster, 41
embryonic rotation, 131
En, 89
endoderm, 4, 63–68
  lineage, 63
  specification mechanisms, 64–66
  differentiation mechanisms, 68
endodermal strand, 63
endostyle, 1, 18, 63, 179, 184–186
enhancer trapping, 38
Enterogona, 3
enteropneust, 175
ENU-based mutagenic screen, 36
ependymal (glial) cell, 11,
    89, 101
Eph, 65, 127
Ephrin, 119
ephrin-Ad, 110, 116, 126
epibody, 70
epicardial budding, 168
epicardial epithelium, 168
epidermal compartmentalization, 72
epidermal sensory neuron, 101
epidermis, 4, 69–75
  lineage, 69
  specification mechanisms, 70
  differentiation mechanisms, 71,
    72
ERK, 20, 59, 155–157
  ERK1/ERK2, 20, 59

ERK3/ERK4, 20,
  ERK5, 20,
esophagus, 63
E-twenty six (Ets), 22
  Ets1/2, 121, 140–142
euchromatic sequence, 19
evolutionary constraint, 77
experimental manipulation (of embryonic
    cells), 31
expressed sequence tags (ESTs), 19, 26
extension (of notochord), 79, 84
ezrin–radixin–moesin (ERM), 85, 86

factor B (Bf), 160
fate map, 114
fertilization, 42, 145, 152–155
fester, 172
FGF (signaling pathways), 55, 92, 100, 119,
    183, 190
  FGF/MAPK, 129
  FGF/MAPK/Ets, 141
  FGF/MEK/ERK, 92
FGF8/17/18, 89, 94, 95
FGF9/16/20, 59–61, 65, 71, 72, 81, 91, 101,
    109, 111, 115, 116, 126, 127, 140, 142
fibrinogen-like (Ci-fibrn), 84
ficolin, 160
filter feeder, 184
first heart field (FHF), 143
fish/tadpole-like larvae, 179, 183
5S rDNA, 26
floor plate, 82, 94
Fog (friend of GATA), 70, 115
follicle cell, 9, 150
follicle stimulating hormone (FSH), 150
follicular cell, 150
follicular thyroid, 184
forkhead box (Fox), 22, 126
  FoxA, 65, 81, 82, 115, 186
  FoxAa, 121, 126
  FoxB, 94, 100
  FoxC, 84
  FoxD, 65, 81, 82, 111, 115, 126, 127,
    184
  FoxE4, 186
  FoxF, 27, 127, 141
  FoxHa, 100
  FoxQ1, 186
forward genetics, 35, 38
Fos-a, 83
free-living ancestor, 175
Frizzled1/2/7, 83
Frizzled3/6, 65

Fusibility(Fu)/HC, 172, 173
fusion (reaction), 172

γ-aminobutyric acid (GABA), 99
    $GABA_B$, 99
GABA/glycine receptors, 99, 100
    $GABA_B$ *R1*, 99
    $GABA_B$ *R2*, 100
GABA/glycine transporter (VGAT), 99
GABA/glycinergic neuron, 99
gametogenesis, 42, 145
gastrulation, 11, 121
GATAa, 70, 84, 91, 115, 141, 147
GCNF, 147
gene duplication, 20, 176
    genome-wide duplication, 20, 176
gene interference, 33
gene regulatory network (GRN), 7, 57, 122,
    126
genetic cascade, 79
genetics, 35
genome
    C. intestinalis, 19–26
    C. savignyi, 22
germ cell, 145–148, 169, 171
germ layer specification, 116
germ line, 36, 145, 171
    precursor cell, 171
    transgenesis, 36
germinal vesicle (GV), 150
    germinal vesicle breakdown (GVBD), 42,
        150
germplasm, 145
Ghost database (website), 27
gill slits, 1, 107
glandular zones (of endostyle), 186
glutamatergic neurons, 97
gonad, 1
    formation, 171
gonadotropin-releasing hormone (GnRH), 150
granular hemocyte, 162
granulocyte, 162
green fluorescent protein (GFP), 32

hagfish, 176
Hairy/Enhancer-of-Split, 133
Halocynthia roretzi, 4, 9, 17, 22, 31, 45, 46,
    49, 50, 55, 56, 59, 60, 61, 65, 67, 70, 72,
    73, 82, 100, 107, 108, 131, 145, 148, 155,
    160, 172, 188, 190
Hand (HAND), 118, 127, 141
haploid sperm hypothesis, 153
heart, 1, 137–143
    differentiation mechanisms, 141–143
    lineage, 138–140
    specification mechanisms, 140,141
hedgehog (signaling), 23, 94
hemoblasts, 162, 168
hemocytes, 162
heterochrony, 175
heterozygosity, 22
Hex, 127
Hif, 100
high mobility group (HMG), 22
histone, 26, 131
    genes, 26
    modification, 131
HNF-3β/forkhead, 94
homeobox family proteins, 22
    Hox gene cluster, 20, 34, 127, 129
    Hox1, 89, 91, 94, 129
    Hox2–4, 34, 129
    Hox5–6, 129
    Hox7, 20, 127
    Hox8, 20, 127
    Hox9, 20, 127
    Hox10, 129
    Hox11, 127
    Hox12, 127, 129
    Hox13, 127
    Hox11/12/13, 34
Homo sapiens, 19, 132, 177
horizontal gene transfer, 20, 73
horseradish peroxidase (HRP), 32
house (of larvaceans), 75, 188
HrUrabin, 155
HrVC70, 155
HSP40, 148
hyaline amebocyte, 162
hydrostatic skeleton, 79
hypothalamus, 150
    hypothalamic–pituitary–gonadal axis (HPG
        axis), 150
    hypothalamus–pituitary–gonadal–liver axis
        (HPGL-axis), 150

immunoprecipitation, 75
incurrent oral (branchial) siphon, 1
induction, 59, 71, 118
innate immunity, 159–162
inner tunic, 69
intercalation (of notochord), 79, 85
intestine, 1, 63
invariant lineage, 13, 119
iodine, 1
    iodine-binding zones (of endostyle), 186

IQGAP, 84
I-SceI mediated transgenesis, 36
Islet, 95, 143

jaw, 2
JUN N-terminal kinase (JNK), 20, 155–157
juvenile, 16, 17

Kaede, 32, 101
KH gene model, 19
Klf15, 83
Kowalevsky, Alexander, 5

lacZ, 33, 34
lam3/4/5, 84
lamellipodia, 85
lamprey, 83, 176
   larval lampreys (ammocoetes), 184
lancelets, 2
larvaceans, 3
larval morphology, 11–13
left–right (L–R) axis, 9, 113, 129–131
lefty/antivin, 65
leprecan, 83
Lhx, 121
   Lhx-3, 65, 67, 121, 140
life cycle, 41
lineage, 5, 13–16, 61, 63, 69, 91
Linneaeus, Carl, 4
localized maternal mRNA, 55
luteinizing hormone (LH), 150
lymphocyte-like cell (LLC), 162

macho-1, 46, 53–56, 109, 115, 121, 127, 140, 146, 189
major histocompatibility complex (MHC), 159, 172
mannan-binding lectin (MBL), 160
   MBL-associated serine protease (MASP), 160
Manx, 190
marker lines, 38
mass spectrometric (MS) analysis, 28
maternal factor, 53, 65, 70, 115, 119
maternal gene cluster, 41
mediolateral region, 71, 84, 94
   patterning, 94
   polarization, 84
Meis, 100
MEK/ERK signaling pathway, 81, 82, 91
mesenchyme, 4, 13, 107–111
   lineage, 107–109
   specification mechanisms, 109–111

Mesp, 27, 121, 127, 138, 140
metamorphosis, 16–18, 101, 145, 155,156
MEX-3, 49
midbrain–hindbrain boundary (MHB), 89, 91
   MHB organizer, 91
MHC class I and II genes, 159
micro RNAs (miRNAs), 133, 134
microarray, 35, 82, 95, 100, 157
microtubule, 85
midline region (of epidermis), 71
Minos, 36–39
miR-124, 133, 134
misshapen, 83
Mitf, 184
mitochondria, 42, 54
mitogen-activated protein kinase (MAPK), 20, 155
   MAPK signaling, 110, 155
Molgula occulta, 189, 190
Molgula oculata, 4, 189, 190
Molgula tectiformis, 189, 190
Molgulidae, 4
monocilia, 131
monoclinic I$_\beta$ allomorphs (of cellulose), 75
Morgan, Thomas H, 5
morula cell, 162
mosaic development, 119
motor ganglion see visceral ganglion
motor neuron, 95
mouth, 63
msxb, 127
Musashi, 38
muscle, 4, 13, 53–62, 107, 137–143, 179
   adult body-wall muscle, 107
   adult siphon muscle, 107
   juvenile heart muscle, 137–143
   larval tail muscle, 53–62
      differentiation mechanisms, 55–58
      lineage, 53, 59, 61
      primary (B-line) muscle, 53–58
      secondary (A- and b-lines), 59–62
      specification mechanisms, 54–55, 59–62
muscle-type actin, 179
Mus musculus, 19,
mutagenesis, 35, 36–39
Myc, 171
myocardial tube, 137
MyoD, 55, 58
myofilament, 55
myoplasm, 9, 42, 53
myosin, 57, 58, 84
   myosin binding protein, 58

myosin, (*Continued*)
   myosin heavy chain, 57, 58
   myosin heavy chain 1 (MHC1), 58
   myosin Iβ, 84
   myosin light chain alkali, 84
   myosin regulatory light chain, 57, 58

NADPH, 162
nanos, 171
NCAM, 84
neck (region), 11, 89, 91
necrosis, 172
Nemo-like kinase, 83
neoteny, 175
nerve cord, 11, 89, 99
nervous system, 89–102, 181
   adult nervous system, 102–104
      differentiation mechanisms, 97–100
      larval central nervous system, 89–102
      larval peripheral nervous system, 100, 101
        lineage, 89–91
      specification mechanisms, 91–97
netrin, 84
neural crest, 184
neural gland, 1, 18, 89, 102
neural plate, 11, 91, 95
   morphogenetic movement, 95
neural tube, 2, 94, 95, 181–183
Neuralized, 133
neurohypophysial duct, 91, 101
neurons, 11, 89, 95–102
neurotransmitter, 97–100
neurulation, 11, 94
NFAT5, 83
NF-κB, 160
NK4, 127, 141
Nk6, 95
Nkx2.1, 100
Nodal signaling, 59–61, 71, 82, 92, 95, 109, 118, 119, 131, 183, 190
   Nodal, 60, 61, 72, 91, 126, 129
nonfusion reaction (NFR), 172
nonmuscle myosin II, 85
Not, 65, 67
Notch signaling, 23, 82
   Notch, 82, 91, 92, 133, 190
   Notch/Delta, 82, 119, 121
notochord, 2, 4, 13, 77–87, 179, 183
   differentiation mechanisms, 82–87
   lineage, 77–79
   specification mechanisms, 79–82
notochord sheath, 83
NoTrlc, 111, 141, 127

Nramp (natural resistance-associated macrophage protein)/DCT1 family, 164
nuage granules, 147
nuclear localization, 65
nuclear receptor, 22

ocellus, 13, 100
Oikopleura dioica, 75, 176, 179, 186
Olfactores, 3, 175, 183
oogenesis, 42, 150–152
ooplasmic segregation (movement), 9, 42
oozooid, 167
open circulatory system, 137
operons, 27
Ortolani, Giuseppina, 5
Oryzias latipes, 151
otolith, 13, 100
Otp, 100
Otx, 70, 73, 89, 91, 94, 121
outcurrent atrial siphon, 1
outer tunic, 69
over expression of genes, 32

papilla (palps), 11, 16, 72, 155
PAR-3, 44
PAR-6, 44
paraxial mesoderm, 84
pathogen-associated molecular patterns (PAMPs), 159
Pax2/5/8, 89, 94, 186
PCP signaling pathway, 84, 85
PDK2, 153
pedomorphosis, 175
PER-like, 109
peribranchial budding, 168
peribranchial epithelium, 168
pericardial coelom, 137
peripheral nervous system (PNS), 4, 13, 71, 89, 100, 101
perivitelline space, 150
perlecan, 83
Perophora, 168, 169, 188
pet-1, 50
pet-2, 50
pet-3, 50
Pgc, 147
Phallusia mammillata, 4, 22, 31, 46, 145, 190
pharynx, 1, 18
Phlebobranchia, 4
photoreceptor cell, 100
Phox2, 101
phylogenetic footprinting, 34
PIE-1, 147

pigment cells, 13, 100
Pitx, 73, 95
Piwi, 171
Pleurogona, 3
polar body, 9, 114
Polyandrocarpa, 167
Polyandrocarpa misakiensis, 169, 171
Polyclinum, 169
polymorphism, 34, 172
POPK-1, 46, 49, 50, 115, 147
postanal tail, 179
posterior brain, 11, 89, 91, 99
posterior vegetal cytoplasm/cortex (PVC), 10, 115
postplasmic/PEM mRNAs, 9, 42–46, 50, 115, 146, 147
  pem, 45, 46, 55, 146, 147
  pem-3, 46, 49, 147
  Type I postplasmic/PEM RNAs, 45
  Type II postplasmic/PEM RNAs, 46
preoral lobe, 16, 155
Prickle(pk), 83, 95
primary function of *Brachyury*, 183
primordial germ cell (PGCs), 13, 145, 147
propagative budding, 167
prospective juvenile organs (PJOs), 17
protein mass fingerprint (PMF), 28
proteomics, 27–29, 148
protochordates, 5
PTBP1, 134
P-TEFb, 147
p38, 20
pterobranch, 175
Pyuridae, 4

radial spoke, 148
Raldh2 (retinaldehyde dehydrogenase 2), 129
raphe, 137
receptor tyrosine kinase (RTK)/MAP kinase, 23
red fluorescent protein (RFP), 32
redox mechanism of vanadium, 164
regeneration, 103, 104, 167
  nervous system, 103, 104
rejection (reaction), 171
repression, 147
restriction of developmental fate, 13
retinoic acid (RA), 127, 129, 169
Reverberi, Giuseppe, 5
Rho/ROCK (Rho-associated kinase), 94
RhoDF, 141
ribonucleic acid polymerase II (RNAPII), 147
Ritterella, 169

RNA-based interference (RNAi), 33
rough cortical endoplasmic reticulum (cER), 42
Roux, Wilhelm, 5
RTK signaling, 140
Rx, 129

Sad-1, 46, 49
salps, 3
SCP1, 134
sea urchins, 64, 160
second ascidian stage, 17
second heart field (SHF), 143
secondary function of *Brachyury*, 183
secondary muscle, 59, 190
secondary notochord, 77, 190
self-incompatibility (SI), 152, 153, 155
  SI factor, 152
  SI locus, 153
self-ligand, 172
self/nonself-discrimination, 152
self-/nonself-recognition system, 159
self-sterility, 5, 152
SELP (P-selectin), 83
seminiferous tube, 148
sensory vesicle, 11, 89, 91, 99, 100
septal mesenchyme, 168
sessile ancestor, 175
sFRP3/4-b, 127
Shankouclava, 186
signaling pathway molecules (SPMs), 23
signet ring cell, 162
Six3/6, 100
Snail, 84, 126
solitary (simple) ascidians, 4, 167
solitary lifestyle, 188
somite, 2, 179
Sonic hedgehog (Shh), 82, 94
SoxB1, 127
sperm, 28, 148
sperm-activating and -attracting factor (SAAF), 152
spermatogenesis, 42, 148
spermiduct, 148
splice leader (SL) trans-splicing, 27
stably expressed gene cluster, 41
STAT, 83
STAT5/6-b, 83, 84
stem-like cells, 102, 167
s-Themis, 152, 153
stigmata, 1, 18
stolonial budding (strobilation), 167–169
stomach, 1, 63, 108

Strongylocentrotus purpuratus, 176
Styela, 188
Styela plicata, 4
supporting zones (of endostyle), 186
survival budding, 167
sushi, 157
swimming juvenile (sj), 38, 73, 157, 188
synteny, 176

tachykinins (TKs), 150
   TK-I, 150
   TK-R, 150
tadpole-type larva, 4
tailbud embryo formation, 11
tail-regression fail (trf), 157
tailless species, 189
tandem repeats, 19
T-box, 23, 55, 58, 115, 121, 140, 143
   Tbx1/10, 143
   Tbx6, 58, 115, 121, 127, 140
      Tbx6b, 55
      Tbx6c, 55
Tc1/*mariner*, 36
T-cell receptor, 159
Tcf, 100
tensin, 83
test cells, 9, 150
testicular epithelium, 148
Thaliacea, 3
thyroid gland, 1
Toll-like receptors (TLRs), 159
Tolloid, 127, 141
TPO, 186
tracing of cell division, 32
transcription activator-like effector nucleases (TALENs), 33
transcription factor (TF), 7, 22
transforming growth factor (TGF)-β, 23
transgenesis, 35, 36
transitory larval organs (TLOs), 17
transposable elements, 19
transposon, 36
   transposase, 36
   transposon-based (-mediated) mutagenesis, 36–38
triclinic Iα (of cellulose), 75
tripartite organization of the CNS, 89
trl1, 160
tropomyosin, 57
troponin, 35, 57, 58
   troponin C, 57
   troponin I, 35, 58

troponin T, 57
trunk lateral cells (TLCs), 13, 17, 107, 109–111
trunk ventral cells (TVCs), 13, 17, 107, 108, 138–141
TTF-1(titf1), 65, 184, 186
tunic, 1, 69, 107, 188
   tunic cells, 107
   tunic formation, 186
Tunicata, 1, 2
tunicin, 1
Twist-like1, 109, 111, 118
Twist-like2, 184
two-color fluorescent in situ hybridization (FISH), 23, 24
two-compartment heart, 142

UG dinucleotide–repetitive elements (UGREs), 50
unicellular striated muscle cell, 55
upstream activating sequence (UAS)–Gal4 system, 38
upstream cascade, 57, 82, 140
   motifs, 57
   signaling cascade, 82
U2 spliceosomes, 27
U12 spliceosomes, 27
Urochordata, 1, 2, 175

vagabond, 36
vanabin, 162
vanadium accumulation, 162
vanadocyte, 162, 163
van Beneden, E., 5
Vasa, 145, 147, 171
ventral trunk epidermis, 141
Vertebrata, 2, 175
vesicular Ach transporter (VACHT), 98
vesicular glutamate transporter (VGLUT), 96
visceral (motor) ganglion, 11, 89, 91, 95–99
visceral nerve, 102
vitelline coat, 9, 150
voltage-sensorcontaining phosphatase (Ci-VSP), 97
Vsx, 95
v-Themis, 152, 153

WBSCR27-like, 109
Whittaker, Richard, 5
whole-mount in situ hybridization (WISH), 26, 45, 57, 82, 122

Wnt signaling, 23
  Wnt-a, 83
  Wnt-5, 46, 49, 50
  Wnt5α, 61, 65, 67
  Wnt/PCP, 83
  Wnt/β-catenin, 183
  Wnt9/14/15, 83
WPRW (Try-Arg-Pro-Try) motif, 46

Xenopus laevis, 31, 41, 50, 83, 113–115, 145

YB-1, 148

Zea mays, 19,
ZF-1, 50, 147
ZicL, 55, 81, 82, 95, 121, 126, 127
ZicN, 65, 67
zinc-finger motif, 23
zipper (zipper/nonmuscle myosin heavy
    chain II), 83
Zn-finger nuclease (ZFN) system, 33